Maintenance Management - Current Challenges, New Developments, and Future Directions

Edited by Germano Lambert-Torres, Erik Leandro Bonaldi and Levy Eli de Lacerda de Oliveira

Published in London, United Kingdom

Maintenance Management - Current Challenges, New Developments, and Future Directions
http://dx.doi.org/10.5772/intechopen.100754
Edited by Germano Lambert-Torres, Erik Leandro Bonaldi and Levy Eli de Lacerda de Oliveira

Contributors
Hikmet Erbiyik, Yury Sokolov, Eric Bechhoefer, Junjun Zheng, Hiroyuki Okamura, Tadashi Dohi, Chaoqun Duan, Lei Song, Věra Pelantová, Tsuguhiro Takahashi

First published in London, United Kingdom, 2023 by IntechOpen
IntechOpen is the global imprint of INTECHOPEN LIMITED, registered in England and Wales, registration number: 11086078, 5 Princes Gate Court, London, SW7 2QJ, United Kingdom

British Library Cataloguing-in-Publication Data
A catalogue record for this book is available from the British Library

Additional hard and PDF copies can be obtained from orders@intechopen.com

Maintenance Management - Current Challenges, New Developments, and Future Directions
Edited by Germano Lambert-Torres, Erik Leandro Bonaldi and Levy Eli de Lacerda de Oliveira
p. cm.
Print ISBN 978-1-80356-479-1
Online ISBN 978-1-80356-480-7
eBook (PDF) ISBN 978-1-80356-481-4

Meet the editors

Germano Lambert-Torres received his Ph.D. in Electrical Engineering from the École Polytechnique de Montreal, Canada, in 1990. Currently, he is the director of research and development at PS Solutions and a member of the Scientific-Technical Council at Gnarus Institute, Brazil. Dr. Lambert-Torres has led research in intelligent systems, management of maintenance, and power system operation for many power industries in Brazil and South America. He has also taught seminars and short courses in many countries worldwide at the Institute of Electrical and Electronics Engineers (IEEE), International Council on Large Electric Systems (CIGRÉ), and International Federation of Automatic Control(IFAC) conferences, universities, and research centers. He has supervised more than 120 MSc and Ph.D. theses and published more than 700 journal and technical conference papers. He is also the author/editor of ten books. Dr. Lambert-Torres is a fellow of the IEEE.

Erik Leandro Bonaldi received his Ph.D. in Electrical Engineering from Itajuba Federal University, Brazil, in 2006. He is the CEO of PS Solutions, where he coordinates dozens of employees producing metallic structures, parts, and devices for industries. The company also manufactures materials based on boiler making and machining. All processes are knowledge-intensive, with a demand for high quality. Dr. Bonaldi is currently the head of the Scientific-Technical Council at Gnarus Institute, Brazil, where he is the coordinator of many research and development projects for Brazilian power and oil industries, verifying aspects of oil quality and producing sensors and systems for automation control. His research interests include maintenance management, industrial electronic processes, automation, predictive maintenance techniques, and artificial intelligence methodologies.

Levy Ely de Lacerda de Oliveira received his Ph.D. in Electrical Engineering from Itajuba Federal University, Brazil, in 2006. He is currently the CTO at PS Solutions, where he develops equipment for predictive maintenance, sensors for many purposes for Industry 4.0, and electronic systems (hardware, firmware, and software) for monitoring electric machinery. He is also a research associate at the Gnarus Institute, Brazil, coordinating developments of data acquisition systems, digital signal processing, and microcontrollers for electronic devices, electronic power systems, power converters, and applications of adaptive and intelligent control in industrial problems. His research interests include DSPs and FPGA developments, condition-based electric machinery maintenance, vibration analysis, and digital signal processing.

Contents

Preface

Modern companies are increasingly trying to reduce costs while maintaining the quality of their services and the time spent producing them. This practice is associated with strict control of maintenance management. To this end, maintenance management information systems have been developed to scan each step of the processes, measure quantities, and plan interventions integrated into the production process to optimize production and maintenance procedures.

The main objective of maintaining maintenance processes is to ensure the quality and continuity of production processes. It should be done by examining all aspects of the manufacturing routine, identifying bottlenecks, and evaluating the criticality of each component or subsystem in the overall manufacturing process of the final product. This management should consider repair costs (or replacement) and miscellaneous expenses, planning the work effectively, with reduced repair times and clear instructions for the procedures to be performed.

Several computational tools for managing maintenance processes help the maintenance manager make more proactive decisions. These tools provide data and performance hints from the various processes, the compilation and analysis of data, trend reports, history of past maintenance, and history of similar equipment (or processes), among others.

Maintenance processes have undergone many changes over time. The first phase was simply corrective maintenance ("repair it when it broke"), in which a component is replaced when it breaks. This type of maintenance causes severe problems for production because the breaks usually occur during the production peaks, thus generating instabilities in quality. The second maintenance phase involves preventive maintenance (also called programmed maintenance), which makes exchanges of components after a specific time. There are two main types of preventive maintenance: time-based maintenance and usage-based maintenance. Time-based preventive maintenance is typically applied to equipment critical in the production process that, once damaged, severely impacts production. In usage-based maintenance, the exchange (or even only one inspection) is done based on production hours or cycles of production. This type of maintenance, which still occurs today in industries, also causes financial problems and unnecessary costs because parts and components can be exchanged in good condition, or unnecessary inspection actions can be performed. The advantage of this type of maintenance process over the previous one is that this process can be programmed, thus avoiding creating problems for production.

Then, with the cheapness of the measuring equipment, there was the possibility of better system observability being used. The next phase of the maintenance process is predictive maintenance. In this process, several quantities are observed to monitor each component's wear. By monitoring the amounts, this type of maintenance mitigates the number of untimely interruptions, reducing unnecessary exchange costs, stops for dispensable inspections, and production disorder.

With the advent of intelligent techniques and the application of higher-order statistical processes, one more step has been taken. Possessing a large number of data and algorithms that enable its interpretation, predictive maintenance is transforming into the maintenance of prognosis. In this last type of maintenance, the fault horizon of a component can be more easily determined according to the quantity and type of its use, always considering the data coming from the measurement system. This maintenance process is aligned with the new guidelines of Industry 4.0, where automation, data exchange, cyber and physical systems, the Internet of Things, and cloud computing are applied.

It is known that there are many maintenance process management strategies and that there is no single formula or a general model for implementing a maintenance process in an industry. Strategies vary depending on the company's size, the maintenance models used, the technological level of the industry, and the degree of observability available in the company. Another aspect that should be considered is the integration of maintenance processes with other processes and areas of the company, such as supply, production, and safety. One way to start or even increase a maintenance process in a company is through the observation and example of other companies with similar characteristics.

This book contributes to the insertion of predictive maintenance in industries and the creation of prognostic maintenance. It includes chapters that address the techniques of managing maintenance processes and presents several topics of interest in the implementation of maintenance, in addition to examples and how they can be applied in industries.

Germano Lambert-Torres, Erik Leandro Bonaldi
and Levy Eli de Lacerda de Oliveira
Gnarus Institute,
Itajuba, Brazil

Section 1

Theory and Main Definitions

Chapter 1

The Maintenance Management

Věra Pelantová

Abstract

The chapter deals with the maintenance management. The review is based on maintenance and management trends in organisations in 2022 and on other findings. There are also historical parallels. Aspects such as maintenance planning and control and management including downtime, resources in terms as material (spare parts and added materials) and personnel are discussed. The issue is linked to other management systems such as quality control, occupational safety, and environment and information security. The methods of planning and control of equipment maintenance are presented. The application of the process approach and the concept of maintenance as a process that needs to be improved are described. The relationship to the Industry 4.0 is mentioned. Linking to risk management is included in this chapter. The chapter is based on a small survey probe in several organisations, and points out identified nonconformities of the maintenance and suggested actions. The goal is effective maintenance for needs of organisations in a current dynamic environment.

Keywords: maintenance, process, management system, organisation, equipment, nonconformity, planning, control, strategy, employee

1. Introduction

This chapter deals with maintenance management in organisations. He submits a literary search in connection with the current situation. Based on the implementation of the probe in several companies, it tries to design effective approaches to planning and managing the maintenance process in the context of a process approach and an integrated management system.

2. Historical review

From the very beginning, when a man made the first product, one began to think about how to prolong its life or restore or improve its function. Various technological procedures have been developed. With the transition from manual to mechanical production, this approach has gained more importance and regularity, as well as technical sophistication. Sophisticated mechanical components and then electronic components have made the equipment a more productive, but on the other hand more complex and vulnerable. From this reason, the probability of his failure increased. That status had to be stopped. It went from maintenance to failure to preventive maintenance. Gradual digitisation has resulted in greater diagnostic development and the birth of predictive maintenance. Now, the maintenance is in the stage of integration with the production into one unit. However, the conditions of the significant environment of organisations and the internal context of their maintenance are also changing. Maintenance management must respect the process approach and its aspects. However, most organisations suffer from inertia. Therefore, this development is yet to come, and organisations face a number of difficulties.

3. Publications and the maintenance management

Equipment outages, human errors and product quality deviations are always a signal for organisations to correct maintenance. They lead to efforts to mitigate the impact on production problems as the article [1] mentioned. Many publications therefore deal with partial and general studies in the scope of this topic maintenance management. For preliminary maintenance planning algorithms for steel companies, the text [2] describes generally the study [3]. A study of tools to support maintenance decisions in discrete production is in the study [4]. It brings the issue of processing a group of tasks in parallel on multiple equipment, as in the article [5]. The preventive maintenance planning model for serial parallel systems is described in the publication [6].

Principles of maintenance are examined with respect to time and cost in the article [7]. The creation of a decision-making system for the maintenance of the spatial arrangement of equipment for the city's road infrastructure is described in the publication [8]. Papers [8, 9] examine the assessment of the level of maturity of a heavy equipment maintenance management system. The solution of barriers and their relations in industrial production are solved in the text [10]. An empirical study of the relationship between maintenance management and employee performance is described in the article [11]. The article [12] devotes to muscle and skeletal disorders in maintenance employees in connection with the risk assessment of these activities. Critical analysis of models that combine maintenance, lean manufacturing and the Industry 4.0, and design of its own for predictive maintenance is mentioned in the article [13]. Multi-target optimisation algorithm for wind turbine maintenance is used in the publication [14]. The explanation in computer-aided maintenance management when considering aspects of the Industry 4.0, such as neural network, models, clouds, the Internet of Things, the article [15] provides. The study [16] assesses and categorises maintenance services across their life cycle and in relation to the Industry 4.0.

Current trends in the production management and the maintenance management are described in texts such as [17–22].

4. Methods

Various management tools are used in the performed studies to streamline maintenance work. The basis is the cooperation of maintenance department with production department in the organisation according to the text [19]. It should be noted that maintenance planning should overlap with production planning, as it is noted in the text [21].

Due to conditions of the substantial surroundings, work with resources and hybrid work [22] are balanced for the predictive maintenance. Emphasis is placed on the installation of sensors [20]. Preliminary maintenance planning can be based on the Genetic algorithm with two-phase optimisation, where the integration of the organisation's strengths in the article [2] takes place. The planning of each cell system in the text [6] is based on the same algorithm and combination of the maintenance after failure and the preventive maintenance in the article [14].

Various characters of the maintenance process are monitored in organisations, such as: MTTR (Mean Time to Repair), MTBF (Mean Time Between Failures) and OEE (Overall Equipment Effectiveness) to solve outage problems [19, 23]. The small OEE is at 55–70% in the article [21]. The determination of the maintenance policy is based primarily on the characters of the number of failures, the number of operating cycles of the equipment and the time of performing the part replacement according to the text [7]. Based on a search of publications and analysis, models are created, including together maintenance, lean production and the Industry 4.0 in the article [13]. Reducing maintenance costs while increasing production performance and assessing several scenarios in the maintenance is described in the publication [14]. An important source of knowledge is also a critical analysis of computer-aided maintenance management systems as in the text [15]. The non-technical sphere brings the interconnection of sustainable, social and economic requirements for technical systems as in the text [16]. Multi-criteria analysis is also used for the maintenance decision-making system. All stakeholders are involved, which strengthens the solution of the problem, which is pointed in the article [8]. The production process depends on production speed, the number of nonconformities, system availability and other performance characters, such as complex KPI metrics. The context of the information can be determined *via* the semantic profile of a part of the system as in publications [1, 18]. The Pareto analysis of maintenance barriers in the organisation's production system is mentioned in the text [10].

The evaluation of equipment criticality is performed through setting priorities and decisions on maintenance with the help of data from computer systems as MES (Manufacturing Execution System) and CMMS (Computerised Maintenance Management System). It is based on cooperation between maintenance and production to increase productivity without increasing investment as in the article [4]. Critical activities are also assessed according to the standard ISO/TR 12295 [12, 24].

To strengthen production performance and to monitor the status of maintenance, meetings and verification of the comprehensibility of tasks by staff in the text [11] are recommended. The health risks of maintenance employees are determined by using an ergonomics study in the publication [12].

The benchmarking in maintenance is promoted as a comparative method. It provides information on the number of unplanned outages and the condition of the equipment. In accordance with the article [23], an audit can be recommended as a tool for measuring the performance of the maintenance process in organisations. Furthermore, it is possible to compare maintenance process through an inventory, where the physical assets and information about them correspond to the data in the computer system as for example CMMS according to the text [18].

Furthermore, methods of evolutionary algorithms, clustering method [25] and/or linear programming [5] are used to plan and schedule maintenance and reduce its costs. Using a digital twin improves the visibility of problems in this process well in the article [17].

CBM methodology is widely used for dynamic maintenance planning, as stated in the publications [3], here in conjunction with the standard ISO 31000 [23, 26]. Methods such as RCM and TPM [23] are also used for a maintenance planning.

Reliability-focused maintenance is recommended as suitable for minimising costs when there is insufficient capacity of qualified staff. The standard ISO 55001 [19, 27] is recommended to enhance sustainability. All maintenance process scenarios over time are also considered, as in the publication [3].

The assessment of the level of maturity of the maintenance process is performed based on the standard ISO 55001 [9, 27].

5. Problems of area

The current maintenance process faces a lot of problems in organisations.

Implications of production planning for maintenance planning have not yet been satisfactorily considered. These two areas are not integrated. There is no suitable approach to rescheduling maintenance according to the current situation in the workshop in the article [2]. There is a lack of analysis of the advantages and disadvantages of various maintenance planning and control algorithms [14] and rigorous data analysis according to the text [20]. There is more interest in a production than in maintenance the text [4] notes. It is necessary to address unplanned maintenance also due to the availability of staff as in the text [25]. A major problem for many organisations is a lack of strategic maintenance planning and overall integration of the maintenance into the strategy of organisations according to the article [23]. The implementation of the Industry 4.0 in the maintenance requires a more sophisticated method of maintenance planning, backed by data and industry knowledge and risk analysis as in the article [3]. Maintenance decision algorithms are often based on discrete state variables [28]. Deviations occur in the communication of sensors, actuators and other devices that can affect production decisions. There is a need to improve the work with contexts and workflows [1]. Problems occur also in the application of predictive maintenance in organisations [17].

Many organisations have difficulty to implement the conception TPM and a spatial arrangement of equipment in a shopfloor as the article [13] describes. The eternal problem is to reduce maintenance costs and increase equipment availability for many articles as in [15, 18]. Problems can be seen in the supply chain in the text [19]. Studies often involve individual facilities or shopfloors, not the entire asset management of organisations. It is necessary to update the methodology for determining the criticality of the equipment regarding new conditions of the substantial surroundings and the consistency between bottlenecks of production and the criticality of equipment. Data for maintenance analysis are often not of excellent quality. Stakeholder requirements are not understood. A static approach is applied, and the holistic concept is not considered. The maintenance is often decided by staff who have any access to it, and even the maintenance staff themselves are not familiar with data analysis [4].

According to the article [10], a common reason for maintenance problems in organisations is a poor communication in organisations and a small interest of management. Employees do not report maintenance problems and do not make improvement suggestions for maintenance. There is a need for more training on maintenance for all employees of the organisation, as texts [18, 23] add. There is a lack of qualified staff [21]. The safety of maintenance staff must be ensured [11] because working conditions of maintenance employees are hazardous to health, as shown by studies [12, 23]. Green technologies are not synchronised with the maintenance and the maintenance is not monitored environmentally [10]. Maintenance workspaces will need to be optimised for energy consumption and a carbon footprint as the article [22] notes.

Overall, problems of maintenance are financial, organisational, environmental, social and technological, as the articles [10, 20] write. The assessment of the level of maturity of the maintenance management system needs to be assessed in relation to the maintenance costs as in the text [9]. Occasionally, there is an inefficient maintenance process, as in the publication [23]. The Benchmarking towards the best maintenance group is problematic from position of internal data, as the authors write [23, 29, 30].

6. Trends

There are a number of trends in maintenance management and in workshop maintenance. The goal of all efforts is to improve the organisation's production performance by more than 40% with the help of preliminary maintenance planning, as stated in the text [2]. The basis is the determination of suitable characters of the maintenance process and their use in sophisticated algorithms that help rapid planning, optimisation and management of maintenance and elimination of staff conflicts in publications [2, 5]. Other factors influencing the maintenance are involved in the prediction, such as vibration [13], energy prices and spare parts wear [14]. It is necessary to establish common characters or metrics for the production process and the maintenance process [4, 16]. Applications of smart technologies such as glasses increase the speed of solving maintenance problems. It also contributed to the transition from the original equipment condition monitoring to continuous multidimensional monitoring and immediate problem solving as it is described by texts [6, 17, 28]. Analysis of the causes of nonconformities will help to subsequently improve the maintenance process [20].

Small- and medium-sized organisations and households need to be supported in the application of computer-assisted maintenance management systems as publications [4, 15] recommended. Data from CMMS are used more often for maintenance planning than from MES, although this one better describes the status quo of the equipment in relation to maintenance and system dynamics. CMMS is used to determine causes of equipment failure, which is a knowledge that production also needs. The management of large volumes of data and their use for planning and managing production and maintenance together is addressed. The task of the future will be to ensure data quality in such a comprehensive management system what the article [4] notes. The programming of the Internet of Things will have to be validated in a real fault environment as in the text [3]. It is necessary to involve expert systems in the maintenance of objects, such as a road condition assessment in the article [8]. Artificial intelligence will affect not only maintenance jobs [19], but also finding compromise solutions between production and maintenance [21].

Frameworks need to be developed for several types of spatial arrangements for maintenance and production [4]. Ergonomic procedures need to be developed, and ergonomic advice should be available to staff. Furthermore, there is a need to expand research into physically demanding jobs that go beyond chronic diseases. There is a need to strengthen the effectiveness of ergonomic assistance in the shopfloor and to create a set of the best ergonomic maintenance procedures, according to the publication [12]. According to the authors [23], approximately 20% of incidents in 2020 year occurred in company maintenance. Therefore, the trend of safety in maintenance in the text [20] is growing.

Organisations should be more committed to a maintenance and risk analysis strategy, according to publications [3, 11]. This must already be included in the design of the equipment. A man and his or her activity cannot be completely excluded from the production process due to manual work, such as maintenance interventions. But his mistake can affect strategy and downtime. Therefore, durable technical equipment and social systems are needed. MES could be a plug-in for an organisation's production planning and management system, for example, according to the article [4]. The maintenance should be part of Advanced Planning Systems according to the text [1].

Furthermore, organisations must follow cultural changes in the substantial surroundings in maintenance, as the text [23] adds. It means creating a culture of teamwork, effective emergency planning and using of the knowledge from a computer support of the maintenance. 'Doing more with a fewer people', according to the article [18]. Human-centred maintenance and the strengthening of communication tools according to the text [22] will help the shopfloor.

7. Discussion

As can be seen from the list of publications above, the maintenance management is an enough broad topic.

The regulating of OEE maintenance metrics as responsibilities does not always seem appropriate for organisations. This method is not yet prescribed by law. However, organisations would need to have some maintenance characters identified and evaluated. The basic characters of maintenance should be a more realistic in the maintenance work environment, which corresponds to the findings in the text [6]. The KPI metric, as stated in the text [18], can be taken in an analogous way. Lack of information on a maintenance performance leads to an inefficient process according to the article [23].

Methods such as RCM, CBM and TPM are used in practice, but in the Czech Republic the RCM method is essential and then the organisations' own approaches.

The maintenance policy determination procedure is described for technicians not only maintenance in the text [7]. This is a helpful solution. It is aimed at professionals, not primarily managers, as is often the case. For example, processes in the article [1] are thought to be technological, but it is necessary to link them with the system according to the process map of the organisation. In addition, planning is not a process, as stated in the text [19], but an activity. Again, this is about dealing with a hierarchical structure and a directive management as in the article [21], which is unsuitable for the maintenance.

Maintenance planning and management algorithms are refined, and the number of iterations is reduced. The period of preventive maintenance is extended, as stated in the article [14]. The algorithms used so far include procedures that perform maintenance at the expense of production time. The article [4] lists several findings that can be agreed with. This is not suitable in terms of shortening the production lead time as the author [31] wrote. Maintenance calendars are often scheduled separately. There may be no link between the maintenance plan and downtime. Production losses should also be a guide for maintenance decisions. The equipment downtime should be a part of inputs to plans and so on. The only time that is crucial for the production is the main technological time. Those interested in streamlining maintenance should focus primarily on problems and waste in the current process and eliminate them.

Teamwork requires the responsibility of all employees for maintenance in the organisation. There is a need to strengthen the image of the need for maintenance in the organisation as a tool for prevention. Qualified staff is required for the maintenance [19]. However, some organisations are not interested in potential new employees. This can subsequently have a demotivating effect on them. The need for a human-machine collaboration is mentioned, but it should have been solved in the past automation efforts. The responsibility or self-responsibility is a good thing for the maintenance process. However, employees must also have other conditions such as rights and resources, including information, which is doubly true in maintenance. All employees in the organisation must be trained in maintenance so that they not only understand its importance, but also ensure their share of maintenance in their own abilities as a part of their activities. Many organisations get into trouble due to the lack of interest of the organisation's management and the lack of sufficient resources for their employees, as shown in the text [23]. Employees must be guided to carry out maintenance tasks effectively, but they should be given an appropriate working environment, as in the text [11]. The fact is that teamwork cannot be required if some form of a hierarchical organisational structure is applied at the same time. This form also leads to other ways of management that are associated with it. The reason the maintenance process is inefficient is the insufficient definition of the process according to its characteristics. Organisations should provide their maintenance staff with sufficient quality protective equipment, which agrees with the authors [23]. However, according to the article [12], the teamwork is also the basis for prevention in maintenance ergonomics.

Surprisingly, the availability of spare parts is not so much discussed in the publications, although the article [14] assesses the field of maintenance quite comprehensively. Maintenance inventory management needs to be based on more suppliers and more extensive forecasts given the current situation, which is in line with the text [20]. There are already mentions of waste management and building efficiency, as in the article [22]. This situation with a lack of spare parts and materials in maintenance (e.g. according to the text [19]) leads to an extension of the equipment's operation in organisations.

From problems described above and according to nonconformities that the author of this chapter encountered, the need for overall integration within the management system is evident. The trend described in the article [19] that environmental and social responsibility issues will be involved in the performance evaluation and the organisation strategy has been deviating since February 2022, although it would be necessary. The holistic of production and maintenance is necessary to increase productivity. This will strengthen the appropriate specification of characters in the integrated management system, in accordance with the text [4]. The problem is that a profit is often required, and the equipment must run constantly. This is a traditional myth of managers. The article [3] describes the involvement of design in risk assessment and maintenance planning. But it is unique. It is associated with production. Therefore, rather than a quality-oriented culture in maintenance, as in the text [23], organisations should build a safety-oriented culture or a holistic culture in the maintenance.

For these reasons, the next text of the chapter focuses on the application of the process approach in the maintenance process and on finding suitable methods for maintenance planning and management.

8. The status quo

The dynamics of the competitive environment is growing [23]. Rapid technological development is underway [19]. Initially, maintenance organisations over the past 5 years have focused on implementing the Industry 4.0 and on the increasing efficiency. The COVID-19 crisis and the current situation, as well as other threats, change the situation of all organisations and affect maintenance as in the text [17]. Maintenance systems come under pressure in such conditions [19]. The supply chains for materials and spare parts are disrupted [20]. As a result of these crises, financial resources are being reduced not only for maintenance [21]. Due to the epidemiological situation, service actions were postponed due to a lack of shift employees and a limit on the number of people who could meet at one workplace. As a result of delayed maintenance, there was a chaining and an increase in the number of problems on the equipment. The root causes of nonconformities have not been addressed [17].

There is a growing need to learn how to work and maintain new complex equipment. Therefore, not only financial and material resources for maintenance are important, but also the staff and their commitment. They are afraid of losing their jobs. Conflicts between maintenance and production are common as in the text [23]. Hybrid work is used [22]. Maintenance employees work under time pressure, under stress. The activities are long-lasting and inconvenient, and performed with an inadequate equipment. The upper limbs and torso are endangered. Handling large loads, working at heights, occasional activities are problematic, maintaining employees age. There is an extended period of training and insufficient knowledge. The organisation of work in enterprises is difficult in the team [12].

Within the survey probe, 10 organisations were examined for the purposes of this chapter on maintenance management. There were seven small- and medium-sized organisations and three large organisations. They operate in the Czech Republic. Their fields of activities are mechanical engineering, electrical industry, automotive industry, glass industry, textile industry and services. The type of production was piece or serial. The organisational structure was full and hierarchical one. There was one group of maintenance employees in organisations. Organisations had implemented basic management systems—quality ISO 9001 [32], occupational safety ISO 45001 [33], environment ISO 14001 [34] and industry standards. They usually do not have implemented the information security management system according to the standard ISO/IEC 27001 [35].

The research interest was focused on the maintenance and its links. Following nonconformities were found out by observations, interviews with companies´staff and by data analysis: outdated equipment, poor storage of spare parts, insufficient identification of facilities, lack of staff, duplication of data, data transcription errors, lack of maintenance records, low material quality, spare parts not available on the market and different maintenance procedures are applied for the same type of work, insufficient staff qualification, work safety incidents, problematic communication, low motivation, hacker attacks and loss of know-how, insufficient training, poor quality previous maintenance work, only one supplier of spare parts and materials for the organisation, data are only collected and analysis is not performed; maintenance plans and production plans diverge. In addition, there is a poor relationship between operators and their equipment.

In terms of the frequency of these nonconformities, the most numerous are occupational safety incidents, incomplete and missing records, communication, different maintenance, and production plans, not performed analyses, reduced material quality and data transcription errors. Due to the COVID crisis, organisations are learning to cope with the lack of maintenance staff and with the size of workshops. The most risk factor in the maintenance is occupational safety incidents. The health and lives of employees are endangered. Hacking attacks are dangerous in terms of data loss, change of instructions or blocking of maintenance work and then stopping the equipment. Making records means consistency and diligence. The situation in supply chains is deteriorating. The prices of items for maintenance are rising. There are not enough of them on the market. Delivery times are too long. The quality of the items is sometimes not good. It is possible to come across fraudulent actions of suppliers.

From the point of view of waste, utilisation of maintenance staff, waiting, unnecessary work and poor-quality material inputs are evident in maintenance according to publications [36, 37].

9. Process approach

The process approach is currently the cornerstone of an organisation's management systems. It is based on the common foundations of the TQM and ISO 9001 [32]. concepts. It is based on decentralisation, cooperation, stakeholder interest, basic documentation of a process, waste elimination, object identification, leadership, communication, value-added solution variability, measurement, comparison, and review as sources of objective evidence and for the continuous improvement of the management system. Effective forms of maintenance also use these interfaces. Some of the types of flexible organisational structure are suitable for the maintenance process. This solution ensures decentralisation, autonomy and initiative. Leadership as a form of leadership supports this. Communication is then free. This makes the whole more flexible. Suitable interpersonal conditions then create an ideal mushroom for creativity and initiative. At the same time, the independence of the individual is supported. At the same time, disinterest and frustration are declining. This also reduces the number of nonconformities in the process. Characters are more apt for such a process. Maintenance procedures can be documented more consistently. Data are collected and analysed as a source for further process development.

10. Planning and control of maintenance

In general, planning in an organisation can be divided into time, material and capacity as in the book [38]. Sometimes, financial planning is added to them. All this can also be applied to maintenance.

The planning tells staff how often to conduct inspections and at what times, for example daily, weekly, monthly. Given the machinery and other assets, the organisation has an idea of how long the maintenance work will take. Normative indicators are still insufficient. Maintenance depends on the design of the equipment and its disassembly options, on the location in the building and accessibility, on the work of previous maintenance. These factors can significantly change maintenance time. Furthermore, it is the detection of the cause of the fault that prolongs the on-going time of maintenance. Maintenance time also increases depending on the technological activities performed.

Material planning tells staff how much material is needed. For example, how many bearings of a certain type on the machine, and how many litters of oil for lubrication. Here, they play a role and are assessed: delivery conditions, availability of the purchased item on the market, its price, failure rate, consumption time, storability, required quantity for maintained objects, frequency of request for use in maintenance. The storage of spare parts and materials must comply with the required storage conditions and the stacking instructions for the item. Due to the current difficult situation of suppliers, it is possible to expect a request for an increase in maintenance stocks. The ABC method can be recommended for the analysis of input materials for maintenance. As with machines, their criticality for the organisation is evaluated here. A novelty in this area will be a greater emphasis on the recycling of materials, their environmental friendliness and the return of packaging.

Financial planning tells staff how much it will cost. These are items such as materials and spare parts, wages and levies, maintenance work, taxes, depreciation.

Capacity planning indicates how much resources are needed for maintenance work. The sources in this case are workers and equipment (machines and hand tools). Qualification and awareness play a role for employees. This category also includes assembly and disassembly procedures and own maintenance procedures, as well as legislative requirements.

There are four specific maintenance levels. They correspond with publications [29, 30].

Equipment inspections can be performed by electronic systems or the human senses of an operator. They are usually performed every day.

The production operator or maintenance employee also performs caring maintenance or service every day. It means replenishing the lubricant and other media that are needed to operate the machine and adjust the machine.

Prevention or repairs to the equipment according to the type of maintenance selected are performed by maintenance personnel over a longer period.

Overhauls of equipment or outages mean that the production process is stopped in a matter of weeks to months. The outage follows mostly preventive maintenance. It is usually necessary to perform maintenance on a larger technical unit or key equipment. This means stopping the production of the company. The shutdown must be planned at all points. All resources must be provided in advance and procedures for maintenance activities, organisational team and documentation, and SW equipment must be prepared. The safety measures of individual objects and the use of protective equipment by maintenance personnel must not be forgotten either. The following must be kept in mind. Residues of substances in equipment can cause safety incidents due to chemical and physical phenomena. During the entire outage, it is necessary to collect data that will be evaluated after the outage. The main risk of downtime is overtime. It could not be foreseen advance. The solution procedure is then determined on the spot, as well as the necessary resources are determined. An extension of the time until the device is put into operation is a consequence. This status creates production losses for the organisation and increases maintenance costs. The evaluation of the shutdown should be reflected in the maintenance documentation. It is an update of the maintenance procedure. An indispensable part of the outage at present is diagnostic devices—vibro, thermo, tribo, measurement of pressure, air leakage, microcrack diagnostics, etc.

The original maintenance planning and management algorithms were based on the business situation in the 1970s. It was a hierarchical organisational structure, directive management, extensive administration, and a complex planning system. However, due to waste, inadequate information and a long flow of data, day-to-day operational interventions had to be used to prosecute both production and maintenance. Maintenance after the failure prevailed. This approach can still be found in some organisations.

PPS (Production Planning and Control) methods are intended for operational planning and management of production. The most common methods such as MRP II, KANBAN, BOA and OPT are available. Maintenance is *de facto* piece to small series production. Therefore, the JIT method is not included. Although the MRP II method is comprehensive in terms of planning and management, its characteristics do not correspond to the process approach. The BOA method is based on the maximum capacity utilisation and the pressure method in terms of production flow, which is also not the best. The comparison with maintenance would therefore be close to the KANBAN and OPT methods, which are based on decentralisation and are based on the pull method. However, regarding the exclusion of time, material and resource planning, they are not sufficient. Therefore, the development of a more suitable PPS method is expected due to the process approach. Therefore, the connection with maintenance planning and management will take some time.

The planning and control management of maintenance must meet goals to improve the cost effectiveness of the maintenance process and increase equipment uptime and eliminate maintenance risks.

RCM (Reliability Centred Maintenance) is a standardised method according to the standard IEC 60300-3-11 [39], which helps implement an organisation's preventive maintenance program. This method considers safety and reduces environmental impact. It determines the technical system, its parts and their functions. It determines the probable causes of failures of the so-called functionally principal elements. Consequences and probabilities of their failures are determined. The consequences are categorised in the decision tree. They are assigned efficient maintenance activities. The result is a maintenance program that can be constantly updated according to the operating situation.

TPM (Total Productive Maintenance) is a comprehensive philosophy of effective preventive maintenance. It is not only focused on the equipment, but also on the involvement of employees who are both production operators and carers for the equipment. Equipment innovations and improvement proposals are put into practice. Staff is trained and emphasis is placed on occupational safety. The performed maintenance must be performed well. *De facto*, this method comprehensively strengthens the culture of the organisation.

RBM (Risk Based Maintenance) is a method that identifies and evaluates the corresponding risks when planning the maintenance of the object. They are assigned a list of faults for which the severity is evaluated. The resulting risk is the product of three parts—human health losses, productivity losses and cost losses. Probabilistic analysis is performed using a fault tree (FTA). From here, the occurrence of faults can be determined in production [40].

FFM (Failure Finding Maintenance) is a method that aims to find hidden faults that are often associated with the security features of the equipment. Even the specification of these maintenance tasks will not prevent equipment failure. The method is based on risk analysis and safety regulations from the manufacturer for the given type of equipment. Maintenance according to this method is performed at regular intervals.

TBM (Time-Based Maintenance) is a method that includes preventive maintenance, performed at regular intervals on a specific device. The goal is to prevent the object or the entire device from failing. The intervals are either time-related or tied to another quantity (e.g. the number of km driven by the car). This method is applied to working equipment.

CBM (Condition-Based Maintenance) is a method based on the identification of physical manifestations of the equipment. No consequence is expected, but manifestations preceding this device failure are detected. It is based on the P-F curve. The point P indicates the detection of the manifestation leading to the fault. The point F indicates a loss of object functionality. The distance between points P and F is the time interval when the maintenance intervention must be performed immediately. In contrast to post-failure maintenance, this type of maintenance can help the organisation prepare for intervention in terms of material preparation, spare parts, tools and maintenance platoon. On the contrary, the method is not suitable if the failure has high variability [29, 30, 40].

The first two of the methods are the most complex and, according to the author's surveys, the most used in organisations.

Maintenance planning and management also depends on the criticality of the object. In this sense, the objects—devices, are classified into three groups:

Key objects are essential for the main production and often complex. They are irreplaceable in technology or performance. They tend to be expensive or new. They are significant due to the depreciation period according to the country.

Common objects can be replaced technologically or numerically. They are moderately complex. Spare parts for these objects are more accessible.

Auxiliary objects are less complex. They are sometimes used for ancillary work. They are usually older.

Each device should have its own passport. This document contains all information about the equipment—production drawings, technological procedures, diagrams, material certificates, test reports, etc. Passport is also the basis on which to determine the criticality of the object and the subsequent planning and management of maintenance.

The calculation of the criticality of the object depends on the cost of failures (1).

$$\textit{Fault costs} = \textit{fault current parameter} * \left[\begin{array}{l} \left(\begin{array}{l} \sum \textit{fault repair costs} + \textit{environmental impact costs} \\ + \textit{occupational safety costs} \end{array} \right) \\ + \textit{mean recovery time} * \textit{hour downtime} \end{array} \right] \quad (1)$$

The second option is to use analysis using point evaluation of a group of characters on the equipment. These features are then evaluated in terms of criticality of this device in a semi-quantitative manner.

For the purposes of this chapter, an analysis of strategic planning and management methods was performed according to publications [30, 31, 41–43]. It was based on the current state of the market and the situation of organisations, considering nonconformities and trends in the field of maintenance management. There are a lot of methods available in this area. However, they are *de facto* modifications of the following methods. The following basic methods were assessed: Porter's Five Forces, Boston Matrix, Balanced Scorecard, Key Performance Indicators, GAP analysis, Management by Objects, MOST Analysis, PESTLE Analysis, Winterling Crisis Matrix, Technique of Scenarios, SPACE Analysis, 10 Megatrends, VRIO Analysis, Forecasting. They are often mentioned as suitable for the strategic management of organisations. However, the focus on maintenance requires a primarily technical concept rather than an economic (market) concept. Furthermore, today's organisation needs to work with risks and security factors. The linking to production must be possible. The method must be able to capture a lot of data from various sources for subsequent analysis and decision making. It should consider all relevant stakeholders. Finally, it must be in accordance with the process approach that is the basis of such organisations from the point of view of management systems.

Some methods have been removed from the menu. They are quite theoretical for practice and their application in business practice would be quite difficult. Other methods are very economically oriented. Others are relevant to different environmental conditions than they currently exist or correspond to a functional approach.

After a thorough analysis of the above methods in terms of their application, algorithm and required data sources, the following are suitable for maintenance management in the current situation: Key Performance Indicators, GAP analysis, PESTLE Analysis, Winterling Crisis Matrix, Technique of Scenarios.

Key Performance Indicators is a performance metric of a process, department or organisation. It includes features of economic, quality, performance and IT services, which helps build the Industry 4.0, features of inventory with respect to spare parts and materials. It fulfils the SMART methodology for goal setting.

GAP analysis is a method of decision making and problem solving in a certain area. It describes the current state, the required goals, determines the difference between the state and the goals, considers nonconformities and measures, proposes, and evaluates solutions. Its safety part is especially important for maintenance. It also has market and legislative parts.

PESTLE Analysis is an analysis of the essential environment of the organisation in terms of strategy. It includes factors such as technical, social, environmental, economic, and political. It also considers nonconformities, events and risks. This method is therefore a risk analysis regarding the internal and external context of the assessed area.

Winterling Crisis Matrix is a risk analysis of the assessed area, which plots the dependence of the probability of risk on its consequences into a matrix.

Technique of Scenarios is an analytical method that devises the course and consequences of various crisis situations with a view to the development of the organisation and changes in the environment. It can also work with qualitative characters (indicators). It then also sets out the procedure for resolving the relevant contingency. This method is a more general.

The Balanced Scorecard method would be borderline for the needs of maintenance management. However, it works less with risks and security.

Smaller organisations can also use the Area Diagram method, where they select and apply the values of maintenance process characters to individual axes. They will then assess them against the target values at certain time intervals.

The widely used SWOT analysis method is more general and is primary intended for the analysis of the organisation's risks. Evaluation frameworks are part of its modifications.

All these selected methods help guide the organisation's maintenance and management efforts in dynamic conditions too. They work with a system of current characters and their target values. From this reason, the strategic planning is applied. These methods help to determine the ways to manage the area so that the organisation can achieve them effectively. This can be called strategic management. The organisation's strategy in the current conditions can be determined with a view to a maximum of 2 years.

11. Characteristics of the maintenance process

Characteristics of the maintenance process is possible solved according to the book [43] and standard as EN 13460 [44], EN 13306 [45], EN 15341 [46], IEC 60300–3-11 [39] and so on.

For most organisations, the maintenance process is the basic process that supports the main production process. Its effectiveness lies in the ability with as few maintenance employees and maintenance equipment as possible to keep as many equipment as possible operational for as long as possible, so as not to reduce function or even to downtime. This is all to happen at minimal cost. The managerial myth of today is that they need to implement predictive maintenance at all costs. As can be seen from practice, for example, the RCM method directly shows that some objects have maintenance after a failure even cheaper. However, aspects such as occupational safety, environment and information security must be considered. If the consequence of the failure affects any of these areas, consistent prevention is in place. This is the case for installations such as the distribution network, water mains and nuclear power plants. Price is not the only aspect. This area shows the essence of a holistic understanding of the integration of management systems in organisations with a significant impact on the maintenance.

Inputs are staff, material and spare parts, equipment, passport of maintenance and its records, medium. Output is equipment under function. Also, there is a waste. Tool can be the Pareto analysis or the FMECA for example. Rules are standard at the end of the chapter. Characters can be for example: cost of maintenance, number of maintenance staff, average maintenance time, number of failures per month, number of safety incidents, average time between failures, delivery time. Owner can be leader of process.

12. Maintenance staff

Employees are an important item of the maintenance. These are not production employees. They must be quite qualified. Their work includes both manual and mental activities. The maintenance team must be able to perform not only mechanical and electrical work, but they must also observe order, cleanliness, degreasing and handle hazardous substances carefully. It must also be able to solve the problem of interconnection of HW (connectivity) and SW for current systems within the Industry 4.0 for proper operation without deadlocks and to ensure quality data transmission. Maintenance employees should have imagination, analytical skills, knowledge of equipment operation, foresight. Therefore, they need a special approach. They must have enough training. Due to the importance and complexity of their work, they must proceed with caution. They need not only responsibilities but also powers and sufficient resources. Their work is, in a sense, creative. Therefore, regarding management schools such as from T. Peters, the standardisation of their work is not appropriate here. It leads to stress and nonconformities. Motivation and communication are important. Timely and full information of employees help speed up maintenance activities, and prevent conflicts and security incidents. The flexible organisational structure is suitable for the maintenance. Equality between members and their mutual trust is essential. Leaders should be their leader, not superiors. A suitable structure is therefore, for example, a team. From a long-term perspective, there is also a need to balance work and personal life. In the past, maintenance employees were called to intervene even when they were out of the emergency status. This had a negative effect on their family life, involvement in associations in the region and their rest. The work pace of the maintenance employees depends on the current work practice, but also on his health condition (e.g. it may be affected by 'the Post Covid') and the type of temperament (e.g. choleric, or melancholic). Appropriate communication contributes to awareness, explanation of unclear issues and encouragement. Ethics, safety and reliability come the first in the maintenance.

13. The Industry 4.0 and maintenance

The Industry 4.0 brings new aspects of maintenance. Compared to the production process, this area is developing more slowly. According to the surveys in which the author participated, the possibility is to state the following. Maintenance work is facilitated by electrical permitting systems and small diagnostic probes for evaluating an object from a multi-character perspective. Maintenance staff use various code readers and data terminals. Furthermore, more self-diagnostics of the equipment is used, which will allow to carefully plan maintenance intervention. The devices are equipped with one or a set of different sensors, which are connected to a SW system for storing and analysing and evaluating data. The equipment can communicate with each other in this way due to production and maintenance. This is the Internet of Things. Auxiliary logistics for maintenance in the form of small autonomous material trucks and autonomous equipment for certain tasks, such as drone inspection of the building, is advantageous. However, this solution is an expensive for some organisations. Internet services are now primarily cloud storage. It is suitable for large volumes of data such the maintenance has. However, it is important where the storage is physically located for security reasons, such as incidents such as theft, damage or data blocking. The organisation's knowledge would be compromised, or the production equipment could be reset or shut down.

Artificial intelligence includes self-learning systems to enhance equipment automation as well as evaluate the life of its spare parts. Businesses collect data from a variety of business areas, including maintenance. They can be marked as the big data due to their volume. However, the data are often incomplete. This in turn leads to the problem that there is nothing to make maintenance predictions. The link between equipment development and maintenance and parts recycling is just beginning to gain ground. Reverse engineering with 3D parts scans and their subsequent conversion into a model and then to 3D printing make it easy to supply some spare parts right now. It depends on the material used and other properties of the unit that the spare part made in this way is reliable for use in the equipment in terms of operation and useful properties. This in turn affects maintenance planning and speeds up its execution.

14. Conclusion

Maintenance planning and management is based on the organisation's current equipment and its capacity, especially the staffing team. It should be noted, however, that maintenance is decided much earlier. This time is the design stage of the device. The design and manufacture of equipment must be designed to allow their subsequent maintenance. This also affects the planning and management of the maintenance process. Dismantled joints, such as welding and gluing, are disadvantageous in this respect. The production must also comply with all relevant standards. Cost savings often lead to material savings, even in areas where wear and tear occur faster due to under sizing. The choice of materials is also important, both in terms of technology and use, and in terms of recycling. Emphasis should be placed on the maintenance stage, called care—service. This applies to refilling lubricants and changing filters. For each equipment, it is necessary to determine whether maintenance is economically worthwhile. Parts should have approximately the same service life so that one component does not burden the equipment that is otherwise successful. Simply simpler construction, even in current conditions, simplifies maintenance and streamlines production. This includes the concept of maintenance as a process and helps to integrate management systems. Ergonomic and safety conditions must be considered. Staff must ensure that maintenance is documented in all activities in a uniform manner in accordance with the maintenance process guidelines. It is necessary to check the stock for the maintenance.

Greater emphasis must be placed on the soft characters of the maintenance process, which also needs to be evaluated. The whole management system is then more flexible. The maturity assessment of the maintenance management system should also be based on this. The recommended planning and management methods then contribute both to the maintenance and to the organisation.

This is a different view of maintenance. Risk management is integrated in it. Within this chapter, only methodological steps were outlined on how to link operational planning and maintenance management and how to proceed in the strategy. Research needs to continue. However, these partial findings are already working in practice. The goal is an efficient maintenance process in organisations.

Acknowledgements

This work was supported with institutional support for long term strategic development of the Ministry of Education, Youth and Sports of the Czech Republic.

DOI: http://dx.doi.org/10.5772/intechopen.105705

Author details

Věra Pelantová
Technical University of Liberec, Liberec, The Czech Republic

*Address all correspondence to: vera.pelantova@tul.cz

References

[1] Koch S et al. Tackling problems on maintenance and evolution in industry 4.0 scenarios using a distributed architecture. In: Götz S, Linsbauer L, Schaefer I, Wortmann A, editors. Software Engineering. Satellite Events, Lecture Notes in Informatics (LNI). Vol. 2814. Bonn: Gesellschaft für Informatic; 2021. Available from: http://www.ceur-ws.org/Vol-2814/short-A5-4.pdf

[2] Qin W et al. Sustainable service-oriented equipment maintenance management of steel enterprises using a two-stage optimization approach. Robotics and Computer-Integrated Manufacturing. 2022;**75**:102311. DOI: 10.1016/j.rcim.2021.102311. Available from: https://www.sciencedirect.com/science/article/pii/S0736584521001915

[3] Martínez-Galán Fernández P et al. Dynamic risk assessment for CBM-based adaptation of maintenance planning. Reliability Engineering & System Safety. 2022;**223**:108359. DOI: 10.1016/j.ress.2022.108359. Available from: https://www.sciencedirect.com/science/article/abs/pii/S0951832022000382?casa_token=pEqh_ZVFPLYAAAAA:VugVfoGHDrmaEa5eSG4bUnF61XR6wz3r7EbHMC4IHrfMz63DtWAAW65WpZasdClVMkvCVr77yg

[4] Gopalakrishnan M et al. Data-driven machine criticality assessment – Maintenance decision support for increased productivity. Production Planning and Control, The Management of Operations. 2022;**33**(1):1-19. DOI: 10.1080/09537287.2022.1817601. Available from: https://www.tandfonline.com/doi/full/10.1080/09537287.2020.1817601

[5] Dolorme M et al. Solution methods for scheduling problems with sequence-dependent deterioration and maintenance events. European Journal of Operational Research. 2021;**295**(3):823-837. DOI: 10.1016/j.ejor.2021.03.067. Available from: https://www.sciencedirect.com/science/article/pii/S0377221721003726

[6] Tavassoli LS et al. A new multiobjective time-cost trade-off for scheduling maintenance problem in a series-parallel system. Mathematical Problems in Engineering, 2021, 5583125, 2021, pp. 1-13. DOI: 10.1155/2021/5583125. Available from: https://econpapers.repec.org/article/hinjnlmpe/5583125.htm

[7] Mizutami S et al. WIB (which-is-better) problems in maintenance reliability policies. In: Handbook of Advanced Performability Engineering. Cham: Springer; 2021. pp. 523-547. Available from: https://link.springer.com/chapter/10.1007/978-3-030-55732-4_23

[8] Pamukovič JK et al. A sustainable approach for the maintenance of asphalt pavement construction. Sustainability. 2021;**13**(1):109. DOI: 10.3390/su13010109. Available from: https://www.mdpi.com/2071-1050/13/1/109

[9] Siswantoro N et al. The evaluation of maturity level on heavy equipment maintenance management according to ISO 55001:2014. IOP Conference Series: Earth and Environmental Science. 2021;**972**:012033. Available from: https://iopscience.iop.org/article/10.1088/1755-1315/972/1/012033

[10] Karuppiah K et al. On sustainable predictive maintenance: Exploration of key barriers using an integrated approach. Sustainable Production and Consumption. 2021;**27**:1537-1553.

DOI: 10.1016/j.spc.2021.03.023. Available from: https://www.sciencedirect.com/science/article/abs/pii/S2352550921000968?casa_token=Fau1vn8-DjEAAAAA:ZdSMnXzevtq_qQeDEkdhPAamns3F3iptFqDutC5Vl72v4nUbDiBolmFlN0KkhDi8_IppUPwqUw#!

[11] Ntshebe S et al. Facility maintenance management and its effects on employee performance: A positivist approach. International Journal of Higher Education. 2022;**11**(7):47. DOI: 10.5430/jihe.v11n7p47. Available from: https://www.researchgate.net/publication/358231667_Facility_Maintenance_Management_and_Its_Effects_on_Employee_Performance_A_Positivist_Approach

[12] Capodaglio EM. Participatory ergonomics for the reduction of musculoskeletal exposure of maintenance workers. International Journal of Occupational Safety and Ergonomics. 2022;**28**(1):376-386. DOI: 10.1080/10803548.2020.1761670. Available from: https://www.tandfonline.com/doi/full/10.1080/10803548.2020.1761670

[13] Mendes DSFT et al. Proposal for a maintenance management system based on the philosophy and industry 4.0. Revista Producão E Ďesenvolvimento. 2022;**8**(1):e587. DOI: 10.32358/rpd.2022.v8.587. Available from: https://revistas.cefet-rj.br/index.php/producaoedesenvolvimento/article/view/587

[14] Peinado Gonzalo A et al. Optimal maintenance management of offshore wind turbines by minimizing the cost. Sustainable Energy Technologies and Assessments. 2022;**52**:102230. DOI: 10.1016/j.seta.2022.102230. Available from: https://www.sciencedirect.com/science/article/pii/S221313882200282X

[15] Velmurugan K et al. Smart maintenance management approach: Critical review of present practices and future trends in SMEs 4.0. Materials Today: Proceedings. 2022;**62**(6):2988-2995. DOI: 10.1016/j.matpr.2022.02.622. Available from: https://www.sciencedirect.com/science/article/pii/S2214785322012640?casa_token=OyeWjINRnSkAAAAA:M0Aaabkax7V51VqmerO6WDmEgh015Oi7bhed9YKYbEVo0kmCSUFvY_gTzTilzR-BuV3QtFcNIw

[16] Hien NN et al. An overview of Industry 4.0 applications for advanced maintenance services. Procedia Computer Science. 2022;**200**:803-810. DOI: 10.1016/j.procs.2022.01277. Available from: https://www.sciencedirect.com/science/article/pii/S1877050922002861

[17] Authors. Maintenance management trends in 2022. MaintenanceCare.com. 2022;**1**:1-5. Available from: https://blog.maintenancecare.com/maintenance-management-trends-in-2022

[18] Steyn S. 3 big challenges faced by Maintenance Managers in North America in 2021, and how to overcome them in 2022. Maintenance Optimization. AddEnergy. no. 2021;**12**:1-6. Available from: https://blog.addenergy.no/3-big-challenges-faced-by-maintenance-managers-in-north-america-in-2021-and-how-to-overcome-them-in-2022

[19] Authors. Software to Optimize Maintenance and Operations. Sigga Technologies.com; EMEA Region, Schilde, Belgium; 2022. Available from: https://www.sigga.com

[20] Authors. The Technology-Driven Leader in Outsourced Maintenance. ATS.com; Advanced Technology Services, Inc., Kettering, UK; 2022. Available from: https://www.advancedtech.com

[21] Authors. System for Reliable Operation of Equipment. Ukraine: SmartEAM; 2022. Available from: https://www.smart-eam.com

[22] Da Costa FÁ et al. Why HVAC Maintenance Is Necessary in 2022. InfraSpeak.com; Porto, Portugal; 2021. Available from: https://www.blog.infraspeak.com

[23] Phogat S, Gupta AK. Identification of problems in maintenance operations and comparison with manufacturing operations: A review. Journal of Quality in Maintenance Engineering. 2017;**23**(2):226-238. DOI: 10.1108/JQME-06-2016-0027. Available from: https://www.emerald.com/insight/content/doi/10.1108/JQME-06-2016-0027/full/html

[24] ISO/TR 12295:2014 Ergonomics — Application document for International Standards on manual handling (ISO 11228-1, ISO 11228-2, and ISO 11228-3) and evaluation of static working postures (ISO 11226)

[25] Akl AM et al. A joint optimization of strategic workforce planning and preventive maintenance scheduling: A simulation-optimization approach. Reliability Engineering and System Safety. 2022;**219**:108175. DOI: 10.1016/j.ress.2021.108175. Available from: https://www.sciencedirect.com/science/article/abs/pii/S095183202100658X?casa_token=uWBnxi98NfsAAAAA:2lPWoIGOFcm9Pu4wuhtUoHljO9P_qKCwEfa4rglXHXiPQOfz9mva9z-ps_YcxQfxw-e4AvvgDw

[26] ISO 31000 Risk management – Principles and guidelines

[27] ISO 55001 Asset management – Management systems – Requirements

[28] Peng S, Feng QM. Reinforcement learning with Gaussian processes for condition-based maintenance. Computers and Industrial Engineering. 2021;**158**:107321. DOI: 10.1016/j.cie.2021.107321. Available from: https://www.sciencedirect.com/science/article/abs/pii/S0360835221002254?casa_token=-AjZ7VV0-AwAAAAA:tS8xa4oV29jOtp_mLY-RAX5BJVSXB8srtsYpCLl3BUomtQBuQomIV0vCV_buE_dS9BOXm52TQA

[29] Legát V. Údržba zaměřená na bezporuchovost – Reliability centred maintenance. In: ÚDRŽBA ZAMĚŘENÁ NA BEZPORUCHOVOST (RCM), MATERIÁLY ZE XVII. SETKÁNÍ ODBORNÉ SKUPINY PRO SPOLEHLIVOST. ČESKÁ SPOLEČNOST PRO JAKOST, Praha, The Czech Republic; 2004

[30] Legát V et al. Management a inženýrství údržby. Příbram: Professional Publishing; 2016

[31] Lubina J. The Industry Engineering. Liberec: Technical university of Liberec; 2000

[32] Standard ISO 9001 Quality management systems – Requirements

[33] Standard ISO 45001 Occupational health and safety management systems – requirements with instructions for use

[34] Standard ISO 14001 Environmental management systems – Requirements with instructions for use

[35] Standard ISO/IEC 27001 Information technology. Security techniques. Information security management systems. Requirements

[36] Pavelka, M. Naučte se vidět a odstraňovat plýtvání [online]. 2015. Available from: https://www.e-api.cz/25781n-naucte-se-videt-a-odstranovat-plytvani

[37] Svozilová A. Zlepšování podnikových procesů. Praha: Grada; 2011

[38] Synek M. Manažerská ekonomika. 5., aktualiz. a dopl. vyd. Praha: Grada; 2011

[39] Standard IEC 60300-3-11 Reliability management – Part 3-11: Instructions for use – Fault-tolerant maintenance

[40] Ben-Daya M. Introduction to Maintenance Engineering: Modelling, Optimization and Management [Online]. Hoboken, United Kingdom: John Wiley & Sons Ltd; 2016. Available from: https://onlinelibrary.wiley.com/doi/book/10.1002/9781118926581

[41] Authors. Strategic management. In: Management Mania's Series of Management. Managementmania.com; Praha, The Czech Republic; 2022. Available from: https://managementmania.com/cs/strategicke-rizeni

[42] Hupjé E. 9 Types of Maintenance: How to Choose the Right Maintenance Strategy. R2 Relisability.com. The Netherlands: R2 Reliability Pty Ltd.; 2021. Available from: https://roadtoreliability.com/types-of-maintenance/#h-time-based-maintenance-tbm

[43] Pelantová V, Havlíček J. Integrace a systémy Managementu. Liberec: Technical University of Liberec; 2014

[44] EN 13460 Maintenance – Maintenance Documentation

[45] EN 13306 Maintenance – Maintenance Terminology

[46] EN 15341 Maintenance – Key Maintenance Performance Indicators

Chapter 2

Definition of Maintenance and Maintenance Types with Due Care on Preventive Maintenance

Hikmet Erbiyik

Abstract

In this chapter maintenance concept is defined and maintenance types are classified with regard to implementing maintenance policies toward preventive maintenance. Especially achieving planned maintenance policies toward preventive (planned) maintenance and condition based maintenance policies toward predictive maintenance is taken into consideration primarily. Maintenance concept is defined and maintenance types are classified. Due care is given for 'Preventive Maintenance' in this chapter. In general 'Maintenance' term could be defined as; The integration of all possible technical and administrative actions, including planning, supervising, monitoring and controlling toward retaining an item, a system, a machine to restore their original functional state in which they can perform the intended functions. In addition maintenance, include protective and corrective actions to keep the plant operational system in intended conditions or to maintain the acceptable manufacturing conditions. Optimum maintenance policies aims to sustain system reliability and robustness within minimum cost. In line with the progress of industry, increase in the system, material and manpower costs, increasing demand for robustness and the complex structure of the machines increases also the importance of maintenance policies. Maintenance types could be divided into two main parts namely; 1. Preventive Maintenance, 2. Corrective Maintenance. Preventive Maintenance is also classified into following sub groups; 1.a- Planned Maintenance, 1.b- Predictive Maintenance, 1.c- Advanced Maintenance Implementations; 1.c.1 Reliability Centered Maintenance 1.c.2 Risk Based Maintenance. All of these maintenance types elaborated with relevant figures in the chapter. In this chapter Comparison of Planned and Unplanned (corrective) Maintenance (With regard to transaction and output) is defined with a table. Additionally, Comparison of Planned and Unplanned (corrective) Maintenance (With regard to infra structure) is also tabulated. 'Benefits of Preventive & Predictive Maintenance' and 'Predictive Maintenance Methods' are defined with relevant descriptive figures in the chapter. For Corrective Maintenance, basic definitions and corrective maintenance steps, types of corrective Maintenance, improvement strategies in corrective maintenance effectiveness are also given. In the final part of 'Results and conclusions' expected and verified benefits of implementing maintenance policies for planned and predictive maintenance are explained. Comparisons in some maintenance policies is given.

Keywords: preventive maintenance (PM), predictive maintenance (PdM), maintenance policies, corrective maintenance (CM)

1. Introduction

In this chapter maintenance concept is defined and maintenance types are classified with regard to implementing maintenance policies toward preventive maintenance. Especially achieving planned maintenance policies toward preventive maintenance and condition based maintenance policies toward predictive maintenance is taken into consideration primarily.

In the progressing parts of this chapter it is attempted to define that planned maintenance policies and condition based maintenance policies provide beneficial results in terms of overall preventive maintenance.

As the result of this chapter evaluation it is questioned whether to obtain below defined fruitful results via achieving maintenance policies such as;

Maintenance policy ensures that: machinery & equipment are in available and reliable condition, company capable of responding usual and sudden customer demands with regard to utilization of equipment, machinery & equipment is stable and consistent enough to manufacture good quality products, better maintained machinery & equipment is a key for succeeding strong competition, better maintained machinery & equipment does not allow sudden or long standing breakdowns, this result is end up with less inventory loss, higher market share and with better maintained machinery & equipment longer MTBF (Mean Times Between Failures) and shorter MTTR (Mean Time To Repair) scores are obtained, and compliying with JIT (just in time) production approach, better maintained machinery & equipment eases overall cost control [1].

Due care is given for 'Preventive Maintenance' in this chapter. In general 'Maintenance' term could be defined as; The integration of all possible technical and administrative actions, including planning, supervising, monitoring and controlling toward retaining an item, a system, a machine to restore their original functional state in which they can perform the intended functions. In addition maintenance, include protective and corrective actions to keep the plant operational system in intended conditions or to maintain the acceptable manufacturing conditions.

With the final part of this chapter expected and obtained outcomes of maintenance policies are defined with regard to planned and predictive maintenance policies that will cover the overall preventive maintenance benefiits. Comparison is also made between preventive (planned) and predictive (condition based) maintenance policies.

2. Overview of the existing related works

There are various research studies in the recent years on the 'Preventive Maintenance' and Total Productive Maintenance (TPM) issues. In a work by [2] Brankovic Dejan, Milovanovic Zdravko, The Role and Importance of Planning of Maintenance in Industrial Practice, importance of maintenance planning and types of mintenance is explained. In the study of [3] Tran Duc, Dabrovsky Karol, Skrzypek Katarzyna, The Predictive Maintenance Concept in the Maintenance Department of the "Industry 4.0", they have pointed out the importance of predictive maintenance to achieve 'Industry 4.0' and to be competent in the market. In a work of [4] Ötleş S, Çolak, UC, Ötleş O. Artificial Intelligence for Industry, with the aid of machine learning and Internet of Things (IoT) approach predictive maintenance is investigated in order to manage the maintenance management potentials and trends. In another study [5] Paresh Girdhar BEng (Mech. Eng), Girdhar and Associates, Practical Machinery Vibration Analysis and Predictive Maintenance, predictive maintenance techniques, maintenance philosophies, principles of predictive maintenance is explained. In another work [6] Tiena Gustina Amran and Leonardus Sujarto, Early Warning System in Preventive Maintenance as a Solution to Reduce Maintenance Cost, importance of early warning system and The Early Warning System application for computer based Preventive Maintenance implementations is studied. In the work of [7]. Dr. S. J. Lacey, The Role of Vibration Monitoring in Predictive Maintenance, importance of predictive maintenance via advanced vibration monitoring techniques is explained in order to detect the equipment failures before its happening.

In all of these studies different useful aspects of preventive and predictive maintenance implementations are pointed out and in some cases advantages of these advanced maintenance techniques over reactive-corrective maintenance are also mentioned.

3. Methodology

In this chapter main concern is given for preventive maintenance activities. With consideration of recent publications and references a research is made on preventive maintenance. Research methodology depends on the definition of maintenance types and the relevant maintenance policies. Throughout the chapter implementation results of the maintenance policies are questioned. Relevant comparisons are made between maintenance policies. The findings are discussed in the part of 'Results and conclusions'.

4. Definition of maintenance and maintenance types

In general 'Maintenance' term could be defined as; The integration of all possible technical and administrative actions, including planning, supervising, monitoring and controlling toward retaining an item, a system, a machine to restore their original functional state in which they can perform the intended functions [8].

In addition maintenance, include protective and corrective actions to keep the plant operational system in intended conditions or to maintain the acceptable manufacturing conditions. Optimum maintenance policies aims to sustain system reliability and robustness within minimum cost. In line with the progress of industry, increase in the system, material and manpower costs, increasing demand for robustness and the complex structure of the machines increases also the importance of maintenance policies. One of the main reasons in inefficiencies and inconsistincies of production systems is lack of proper maintenance policies or lack of their implementations. However in the recent years, the importance of maintenance policies have been perceived by the industrial, engineering sectors and academic disciplines within operational management and due care have been given [9]. Maintenance types could be divided into two main parts namely; 1. Preventive Maintenance, 2. Corrective Maintenance. Corrective maintenance could be divided into two groups; 2.a- Unplanned repair and Change, 2.b-Foreseen repair and change.

Preventive Maintenance is also classified into following sub groups; 1.a- Planned Maintenance, 1.b- Predictive Maintenance, 1.c- Advanced Maintenance

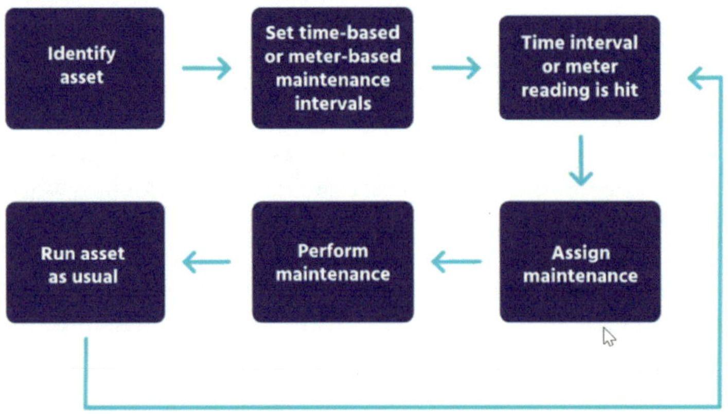

Figure 1.
Work flow of preventive Maintenance [10].

Serial No.	Parameters	Unplanned maintenance	Planned maintenance
01	Failures	High	Low
02	Down time/stoppage	High	Low
03	Product output	Low	High
04	Maintenance costs	High	Low
05	Reliability of equipment/plant	Low	High
06	Availability of equipment/plant	Low	High
07	Percantage usage of equipment/plant	Low	High
08	Spare parts and inventory control	No	Yes
09	Failure Warning / Early Failure Warning	No	Possible

Table 1.
Comparison of Planned and Unplanned (corrective) Maintenance (With regard to transaction and output). (Own compiled).

Implementations; 1.c.1 Reliability Centered Maintenance 1.c.2 Risk Based Maintenance.

A schematic view is given with **Figure 1** for Work Flow of Preventive Maintenance [10].

In **Table 1** Comparison of Planned and Unplanned (corrective) Maintenance (With regard to transaction and output) is given. It is clearly evident from the table that planned maintenance has a wide range of supremacy over the unplanned maintenance with regard to transactions and output.

In **Table 2** Comparison of Planned and Unplanned (corrective) Maintenance (With regard to infrastructure) is given. It is clearly evident from the table that planned maintenance has a wide range of supremacy over the unplanned maintenance with regard to infrastructure. However as it is indicated with (*) marked items planned maintenance would require extra financial allocations.

Serial No.	Parameters	Unplanned maintenance	Planned maintenance
01	Maintenance manpower size	Small	Wide
02	Technical competence of manpower	Low	High
03	Required special equipment	No	Yes*
04	Required expert services	No	Yes*
05	Need for establishment of a special laboratory	No	Yes*
06	Computer back-up is necessary	No	Yes
07	Personel spesific training	No-insufficient	Yes
08	Infrastructure establishment cost	Low	High

**Marked items require extra financial resources.*

Table 2.
Comparison of Planned and Unplanned (corrective) Maintenance (With regard to infra structure). (Own compiled).

5. Preventive maintenance

5.1 Planned maintenance

As a part of preventive maintenance, planned maintenance is to plan the maintenance of machinery, equipment, buildings and plants in advance of their wearing or unplanned stoppage and to prevent them to face breakdown. The time frame of Maintenance plan is usually defined as 6 month or 1 year. Depending on the conditions and structure of the machinery, equipment or plants maintenance intervals defined in weeks, months, quarters or yearly basis. In most cases an initiating advice is taken from machine or equipment manufacturer maintenance manuals for definition of regular maintenance interval. Time based planned maintenance renders the machinery to restore their own original proper functioning state and defers the unexpected breakdowns. During the planned maintenance certain parts of machinery or equipment are replaced with regard to their service life capacity [2].

Benefits of planned (preventive) maintenance policies;

Planned maintenance offers five basic outcomes in general namely;

1. Maintenance cost reduction- By constructing a planned (preventive) maintenance plan small scale failures of the equipment can be detected in advance they turn into bigger problems hence substantial maintenance costs are avoided.

2. Extended Asset Life: Timely and regularly maintaining assets prolongs expected life span and early breakdown in life cycle is avoided.

3. Workplace safety improvement: With the outcome of planned (preventive) maintenance proper functioning of equipment provides also work place safety, hence operators and workers will avoid potential accidents.

4. Improving awereness and company culture: Apart from reducing equipment downtime and failures, planned (preventive) maintenance will also reduce employee absentees due to improved working environment team spirit and moral attitudes.

5. Decreased downtime due to planned (preventive) maintenance: Planned maintenance enables the maintenance team to resolve minor equipment failures before they turn into bigger problems. Due to gathering valuable data on the equipment history with planned maintenance will enable the responsible persons to take preventive actions toward improving life cycle span of the equipment [11]

However in order to obtain above defined benefits, advanced digital infra-structure must be established and a platform of IoT (Internet of Things) must be placed. Futhermore in order to analyze the machinery data, machine learning and forecasting based modeling statistical techniques must be utilized [12].

5.2 Predictive maintenance

As it is understood from the headings, this type of maintenance points out the prediction of breakdown probability of an equipment by automated computurized monitoring and assessment steps and provide a new maintenance plan for failure prevention [3].

For a successfull implementation of predictive maintenance and to ascertain the machinery-equipment proper functioning conditions it is recommended to provide the following data; a. Equipment Operational Records, b. Past data and records on downtimes, breakdowns, performance. c. Condition of machinery-equipment with regard to operating parameters in the past operating period. d. Artificial Intelligence data from machine learning and data analytics. Upon obtaining the above defined data and having benchmarking experience from similar equipment and similar cases a new maintenance schedule is defined [13]. Predictive maintenance provides information on average performance values of machinery and equipment, their potential failures, maintenance state and schedule, the ways how to repair the equipment,..etc. similar data. Hence it provides valuable information in advance of critical production equipment breakdowns for the maintenance team and guides them for the certain repair and maintenance ways and inform them in order to minimize the equipment down time. Due to this beneficial consequence sustainability and efficiency in production is increased and production and maintenance costs is decreased [4].

6. Predictive maintenance methods

Various tools and ways may be employed in the implementation of 'Predictive Maintenance'. Some of the main ways are given below;

a. Failure early warning system

b. Vibration Analysis

c. Thermal analysis-thermography

d. Acoustic Emission

e. Oil & Particle Analysis

f. Corrosion Monitoring

g. Performance Monitoring [5].

6.1 Failure early warning system (IEWS)

One of the prime implementation methods of the predictive maintenance is to utilize the early warning system. Early warning system supports the existing computerized maintenance program. With the implementation of early warning system (IEWS), maintenance activities are conducted more effectively and potential failures of machinery and equipment are detected in advance of failure. With the aid of this system, priorities in maintenance needs of the equipment is ascertained and certain parts that are necessary for replacement, are defined. With the availability of (IEWS) records maintenance costs and potential saving of the organization by avoiding the damage is defined. IEWS provide a valuable maintenance data base for the future use of the organization and for establishing an efficient document control system. As a result with the implementation of IEWS overall effectiveness of equipment and plant usage is increased and it paves the way for further innovative and sustainable maintenance actions [6].

One of the implementation area of an early warning system is using new technology for detecting abnormal equipment performance in power plants. Existing technology can reduce derates and forced shutdowns by providing means to plant operators to adjust-repair small problems before they turn into large problems.

Power generation operators are oriented adopting asset management to improve process efficiency and to increase return on assets (ROA). High value equipment and components such as boilers, turbines, generators and auxiliary systems present an attractive target for asset management since they susceptible to cause derates and forced outages when they fail. Some new technologies in this regard calls predictive condition monitoring, reduces forced outages and derates through actionable early warning of failure of critical power plant equipment. Apart from preventative maintenance implementations, which foreseen maintenance based on failure statistics for a type of equipment problems over time, predictive condition monitoring provides equipment-specific, condition-based early warning [14].

6.2 Vibration analysis

Vibration analysis is a common used predictive maintenance technique that is used to define the existing operating condition of a machinery or equipment in advance of developing problems before they become too critical and might cause unexpected downtime via regular monitoring of equipment vibrations. By way of vibration monitoring, deteriorating or damage of equipment bearings, vanes or blades, belts, mechanical looseness and worn or broken gears,..etc. can be detected [7].

In **Figure 2** [15] Vibration analysis in predictive maintenance is depicted with Early bearing and Gearbox Fault, Late Stage Bearing and Gearbox Fault and Imbalance, Misalignment, Looseness stages.

Basically vibration measurement technique, is an effective, non-intrusive method to monitor machinery or equipment during start-ups, stoppage and regular operation stages. Most frequent usage of vibration measurement is realized on all rotating equipment especially on spindle-bearings, piston-cylinder connections namely various

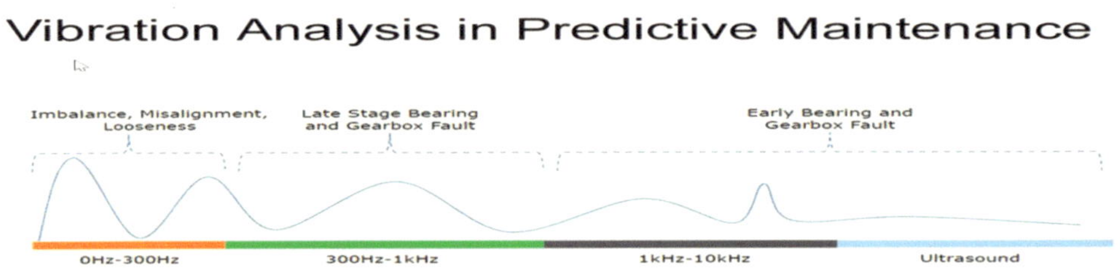

Figure 2.
Vibration analysis in predictive maintenance [15].

types of gas, steam, and wind turbines, compressors, motors, pumps, vantilating fans, rolling mills, gearboxes,...etc. Main parts of vibration analysis system could be defined as; a. Signal pickup(s), b. Signal anlayzer, c. Analysis software, d. A Computer for data analysis and storage. A confugaration can be made among those four basic parts to establish a permanent online system, a periodic analysis system or a multiplexed system that provides sampling of a certain transducers at advance defined time intervals [5].

7. Benefits of vibration analysis

It is possible with vibration analysis to determine the improper cases in machinery and equipment maintenance or repair. Amon those improper practices, improper bearing installation and replacement, inaccurate spindle alignment, or loose rotor balancing can be cited. Early vibration testing renders maintenance staff predictible information on required repairs and necessary parts, enables them take the faulty equipment away from the operation place to prevent any possible hazard, help to prevent equipment ceasing, fosters extending equipment life capability, helps to reduce unexpected equipment breakdowns and failures. Statistically almost over 75% of common revolving equipment failures are related to misalignment and unbalance, hence vibration analysis becomes a prime tool that can be employed to reduce or mitigate repeating equipment failures and problems. As a result a vibration analysis may be employed as prime segment of generic predictive maintenance program [16].

Main components of predictive maintenance in the electrical engineering is defined with **Figure 3** [17].

7.1 Thermal analysis

One of the most effective tools of predictive maintenance is to utilize the thermographic analysis. The technique is based on the infrared thermography (IRT). Since the majority of the machinery equipment failures in the industry becomes evident with temperature changes that can be sensed by regular monitoring with an infrared thermographic system. Thermal data from the equipment is collected via thermal sensor which may be a key source of information for diagnosis and enables the maintenance team to define the failure causes in advance of their occurrence. Hence such an early detection prevents potential future problems before occurrence and high and unexpected repair costs are avoided [18].

Thermal scanning test with thermal camera: Thermal scanning process with thermal camera is made with the aid of infrared beams. All the objects having heat level above −273°C, transmit thermal energy. Visible wavelength that is seen by human

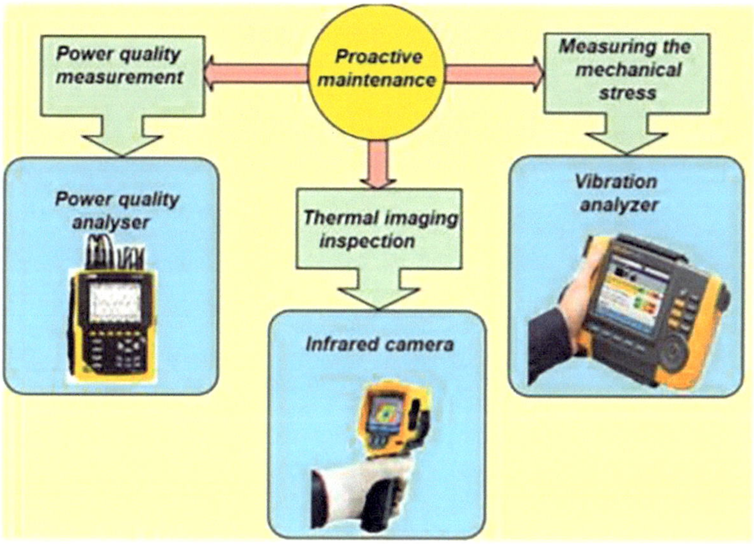

Figure 3.
The main components of predictive maintenance in the electrical engineering [17].

eyes is between 400 nm and 700 nm range. The beams having values below that level; infra violet beams, x-rays or gama rays and the beams having values above that level; infra red, micro wave beams, radio-tv beams can not be seen or sensed by human eyes. Thermal cameras due to self mounted sensors can sense the infra-red beams that are emitted from the hot surfaces of the objects and evaluate them with the aid of a special software and ascertains the temperature values. Heat measurement with thermal cameras will help to define the equipment failures as it was found with the other predictive maintenance techniques. For example on the locations that loose electrical connections are evident, overheating may arise due to increasing resistance or electric motors due to inefficient working conditions may present higher working temperatures than the normal operating temperatures. Due to friction between rotating components unwanted higher temperatures may arise. This excess temperature indicates a loss of efficiency, on the other hand fire may break out unless a preventive measure is taken. Proper functioning of thermal camera will reflect all these problems on the screen directly [19].

Impairment detection with thermal analysis: **Figure 4** shows the thermal camera image for the rotating furnace mantle in cement factory. The aim for thermal analysis is to define the heat distribution in the furnace whether it is uniform or not and to detect the wears of refracter tiles on the mantle interior wall. These analyses are monitored continuously both with mobile thermal camera and on-line thermal scanners. As the result of those analysis life capacity of refracter tiles could be detected and stoppage planning to be made for rotating furnace. Furthermore with the thermal analysis method the following detections could also be made; a. Anomalies in electric panels and connections, b. Failure detection in electric motors, c. Refracter tile wearing on rotating furnace and siclons. d. Leak detection in hydraulic and steam circuits...etc. **Figure 5**.

7.2 Acoustic emission

With the aid of Acoustic Emission method, impairments in the machine elements such as bearings and gearboxes and gas leakages on the pipes are detected. In most

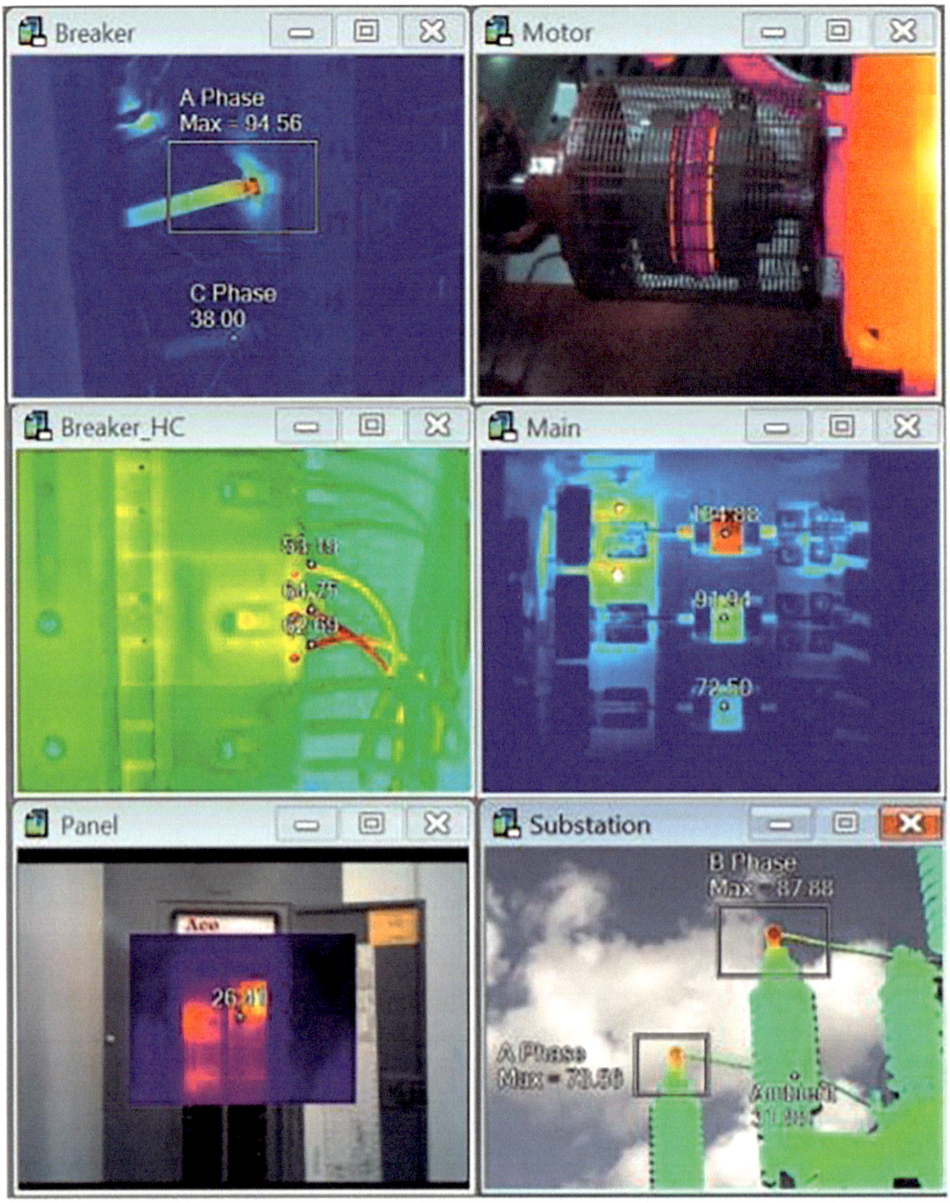

Figure 4.
IR thermal images of an equipment currently used in power electrical installations [17].

cases, in acoustic emission monitoring is made with ultrasonic voice detectors. E.g ultrasonic air leak detector, and bearing voice sensing instruments are utilized for this purpose.

Case study for acoustic emission detection: In a cement factory as it is seen from (**Figure 6**) thickness measurements are made on the cement factory rotating furnace mantle in annual periods and some laminated region is detected on the mantle. For clear definition of laminated region detailed scanning is made with ultrasonic test equipment and laminated area has been detected (**Figure 6**). Upon further review

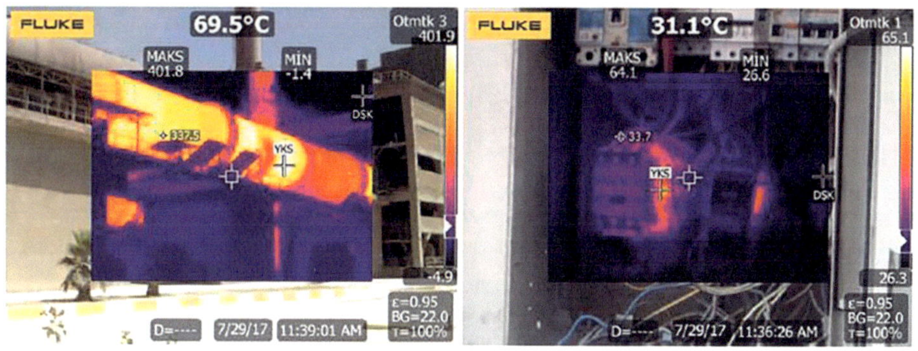

Şekil 7. Döner Fırın Mantosunda Termal Analizi ile Hasar Tespiti

Figure 5.
Impairment detection with thermal analysis on the rotating furnace mantle [19].

Figure 6.
Impairment detection on the cement factory rotating furnace mantle via ultrasonic analysis [19].

and investigations, laminated region has been replaced with the new mantle part. After defining the laminated part on the furnace mantle, the effected region is marked with permanent color marker and it has been monitored and controlled in suitable furnace stoppage periods until the mantle part is replaced. As the result of those controls it is found with the ultrasonic tests that laminated region is propagated and a possible future mantle crack and an unexpected furnace stoppage is prevented with the timely measures.

7.3 Oil and particle analysis

Lubrication is necessary for the mechanical parts of the equipment for smooth operation. With the use of proper lubricant, friction among the mechanical parts is minimized. If the quality of the lubricant (oil) is worsened in the course of time, wear and overheat arises due to excess friction. Lubrication among the moving parts of the equipment is very important. The analysis of lubricant between the contact surfaces, is one of the preferred methods in predictive maintenance [20].

Lubrication in equipment and machinery mechanical parts provides two main benfits; firstly, it provides a preserving film between the moving mechanical parts surfaces, and hence reducing the harmful friction, eliminating unwanted seizing, secondly, lubrication provides cooling of mechanical components, protects the metal surfaces from corrosion, and provides a contaminant deposit free surface [21].

Possible changes in the physical and chemical features of the oil affects the performance characteristics of the lubrication oil, which may lead to performance hampering. That is why it is essential to assess the performance parameters of the oil to clerify that if the oil quality is worsened to a critical level that oil can not fulfill its intended function. There are various lubricant oil evaluating and monitoring techniques that may monitor the oil charactersitics fully or partially. Main causes of the lubricant oil degradation could be attributed to, particle contamination, oxidation and or water contamination [22].

Oxidation products hampers the required viscosity state and lead to wear particles formation that also results additional damage to the mechanical system when they contact with the component surfaces. Wear particles may block the filters and or oil holes and hence causing oil shortage and friction and seizing between moving mechanical components. Worst of all wear particles can tear the filter and high level contamination may occur. Resulting study of wear debris in the oil, enables to detect potential harm in advance of the expected failure so that required preventive measure could be taken [23].

At the result of oil analysis, parameters such as; physical and chemical features of the oil, number and size of the contaminant particules and the pollution of the oil are analyzed in order to make interpretation about the possible future failures. Oil analysis gives us clues about the level or magnitude of impairment on the worn parts. Main reason for the friction of mechanical parts is usually attributed to the low viscosity. Hence viscosity of oil is monitored periodically for assessing overall oil quality, oil is replaced on the point that oil loose its intended features and the equipment is taken into immediated maintenance program [24].

7.4 Corrosion monitoring

One of the important conditional based monitoring method in predictive maintenance is corrosion monitoring. Corrosion is defined as 'metals losing their metallic features by getting into chemical and electrochemical reactions with the surrounding environment'. Corrosion is very important for a country's economy. Recent research indicates that corrosion causes a loss of ¼ of total steel production. Climatic conditions in the cities, rain waters (traces of sulfuric acid or nitric acid), and sea waters are the prime causes of the corrosion.

The most encountered materials type for corrosion are metals since they have a higher tendency with electrochemical reactions. In metals corrosion oxigen is the prime reason. However there are some side effects for corrosion along with oxygen. For example aluminum external surface oxidized very quickly and after surface oxidizing is finished, a resistant protective coating is formed that prevents oxidizing the deeper surfaces. That means external surface is coated with oxygen (corrosion) resistant (Al_2O_3). During corrosion, anodic (electron donating-oxidation) reactions and cathodic (electron receiving-reduction) reactions occur together.

7.4.1 Factors effecting corrosion

Effect of the environment: The rate of corrosion of metals is largely related to the environment in which they are found. The amount of humidity in the environment, acidity-basicity, the ability of air, oxygen or water to pass through the environment, leakage currents and various bacteria appear as initiating and accelerating factors.

Effect of temperature: Increasing ambient temperature increases the rate of corrosion by increasing ion movement. The soil, whose ambient temperature varies between −50 and + 50 0C, freezes at 0 0C and the ion movement speed decreases to a minimum. Increasing the temperature also has the effect of lowering the oxygen concentration. However, this effect is rather weak compared to the reactions caused by increased ion movement.

Effect of material selection: One of the factors that cause corrosion is the use of metals that have potential differences with each other. This is an initiating and accelerating factor of corrosion. For example, stainless steel bolts and gaskets placed on panels made of steel sheet, as a common mistake, cause galvanic corrosion in the area where they are located. In such cases, bolts or gaskets to the main surface should be isolated with plastic.

Differences in properties between grains: As a result of the differences between the grain sizes of the metals and the different concentrations in the two grains, the boundary of the two grains creates a suitable environment for the initiation of corrosion. As a very common mistake is to corrode the welding areas in tanks and similar structures made of stainless steel materials, even though it is not expected by the manufacturer. The way to prevent this corrosion is either to use electrode welding or to apply a galvanic anode cathodic protection system as a preventative.

System design: In systems where corrosive materials are stored, designs should be applied to prevent the accumulation of corrosive medium (water etc.). Also, very thin gaps that can cause liquid accumulation between them should be avoided.

Oxygen concentration of the environment where the system is located: In the same type of soil, the dissolved air concentration may not be the same everywhere. In systems with different ventilation conditions, the system standing next to each other is the anode in one area, while it can act as a cathode in the area next to it, causing electrochemical corrosion.

Effect of soil electrical resistivity: High conductivity in low electrical resistivity regions causes the ionic medium to be more active. Therefore, the corrosion mechanism develops faster.

7.4.2 Corrosion types

Homogeneous (uniform) corrosion: It is the type of corrosion that occurs on the metal surface at an equivalent severity. As a result of corrosion, the metal thickness decreases by the same amount at every point. Metals that are produced from the same type of material in the atmosphere and are not affected by any external factors undergo homogeneous corrosion.

Galvanic corrosion: It is the type of corrosion caused by the use of two materials with different potentials together or the difference in the ground structure. Corrosion caused by the use of different materials creates a galvanic cell between two metals at different potentials when they are in contact with each other, and the active metal acts as the anode and the noble metal acts as the cathode, causing corrosion in the active metal. For example, if copper and steel come into contact, steel will corrode due to copper.

Electrolyte: Solution or moist materials containing ions that conduct electric current. In summary, it is a corrosion phenomenon that occurs on the more electronegative metal surface when two different metals immersed in an electrolyte are in contact with each other.

Galvanic anod: The galvanic anode is the electrode that is used to protect a structure cathodically and that provides current production by dissolving it as a positive ion in the environment. If a more active metal (galvanic anode) is to be attached to a corroding metal, then the electrons required for the cathode reaction are provided by the self-propelled oxidation reaction of the metal connected as the galvanic anode. Thus, all anodic reactions on the protected metal surface are completely stopped. Galvanic anode cathodic protection is also based on this basic principle. In order to cathodically protect a steel pipeline with galvanic anodes, a more active metal (magnesium anode, etc.) is connected to the pipeline. Thus, magnesium becomes the anode in the galvanic battery and the cathode in the steel pipe. Magnesium dissolves at the anode, releasing electrons. These electrons supply the electron requirement of the cathodic reaction. In order for the system to work spontaneously, there must be a potential difference between the anode and cathode enough to overcome the circuit resistance. Types of galvanic anodes. There are three types of galvanic anodes. 1. Magnesium anode, 2. Zinc anode, 3. Aluminum anode.

Crack Corrosion: It is a type of corrosion that occurs in a crack on the metal surface or in a narrow gap. The main cause of this corrosion is oxygen between the crack and the surrounding electrolyte concentration or the difference of metal ion concentration. Since the outer parts of the crack will be the cathode, corrosion does not occur in this region.

Pitting corrosion: Corrosion that occurs in the form of deep and narrow cavities as a result of the concentration of corrosion on very narrow areas is called pitting corrosion. The depth of these pits is approximately the size of its diameter. The mouth areas of the pits are often filled with corrosion products. It is a dangerous local damage with the appearance of tingling on the metal surface.

Stratification corrosion: If intergranular corrosion occurs parallel to the extrusion or rolling surface, it is called stratification corrosion. In this type of corrosion seen in aluminum and its alloys, damage occurs from the grain boundary elongated in the rolling direction. Corroded metal layers are separated from each other and the corrosion products formed cause the material to separate in layers.

Erosion corrosion: In this type of corrosion, which is especially common in pipe systems and ports, the wear rate of the metal increases due to the relative movement between the metal and the corrosive medium. Holes, grooves and trenches form on the metal surface. It manifests itself in many structures in motion in water. The presence of solid particles in the environment further increases the corrosion rate.

Leakage current corrosion: The leakage current of rail vehicles such as trains, trams and subways in the soil causes very severe and rapid corrosion in underground pipes. At every point of the line, there is a current toward the ground. Metal corrodes according to Faraday's Law. In particular, the leakage current emitted from the rail vehicle returns from the pipe to the rail around the point where the negative pole is connected to the rail and creates the risk of corrosion.

Coating failure corrosion: The potential of a coated metal is different from the potential of an uncoated metal. In case of deterioration or perforation of some parts of the coating due to workmanship errors, these regions will become anodes and will corrode. This type of corrosion is a corrosion that concentrates in very small areas on the metal surface [25].

Corrosion monitoring in predictive maintenance is one of the important methods. The damage on the metal bodies or on the metal surfaces realized in different forms as explained in details above. In most cases corrosion in metal surfaces ends up with reduced wall thickness or tears or holes on the metal surfaces [26].

Corrosion mechanism usually defined as an electrochemical process, including charge transfer between anodic and cathodic parts of the system. Corrosion measurement in an electrochemical reaction is usually made by setting up two working electrodes (WE1 and WE2). The current is measured via a zero resistance ammeter. (ZRA). As depicted with (**Figure 7**) [27].

Cathodic Protection (CP) monitoring: One of the most important forms of corrosion protection for submerged/underground structures (such as tanks or pipelines) is cathodic protection. In **Figure 8** [28], a typical cathodic protection architecture is depicted [28].

7.5 Performance monitoring

One of the most important predictive condition monitoring technique is the performance monitoring. In our days high technology performance monitoring instruments

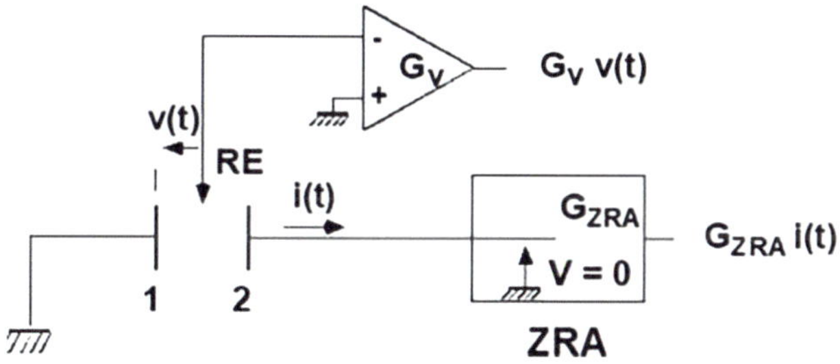

Figure 7.
Practical layout of an EN measurement cell for corrosion monitoring [27].

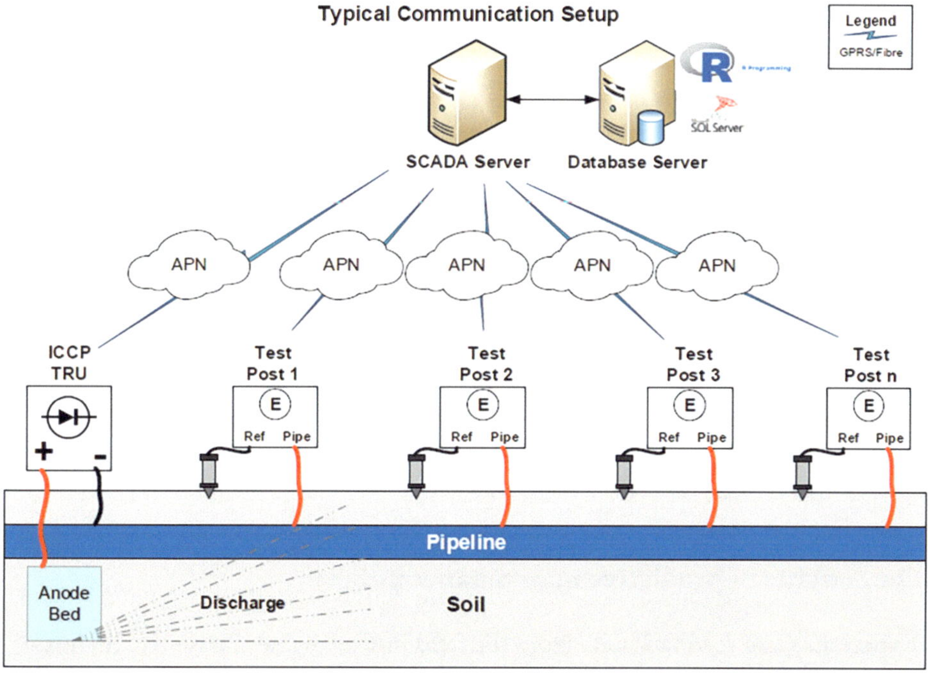

Figure 8.
Practical EN setup using embedded electrodes for corrosion monitoring [28].

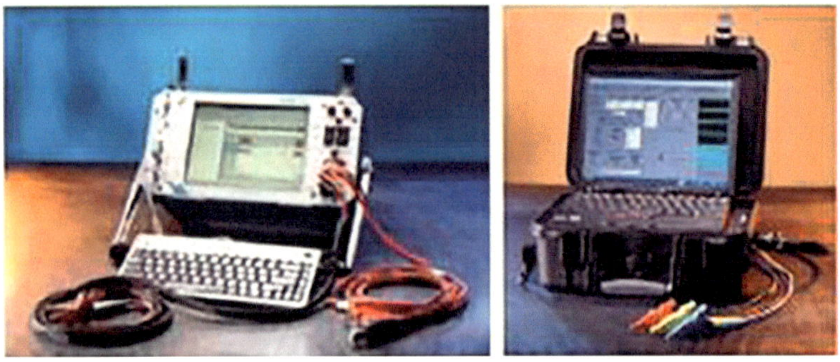

Figure 9.
SKF – Baker AWA IV – Offline Test Instrument SKF – Baker (left), The Explorer – Online Test Instrument (right), for electric motor performance monitoring [29].

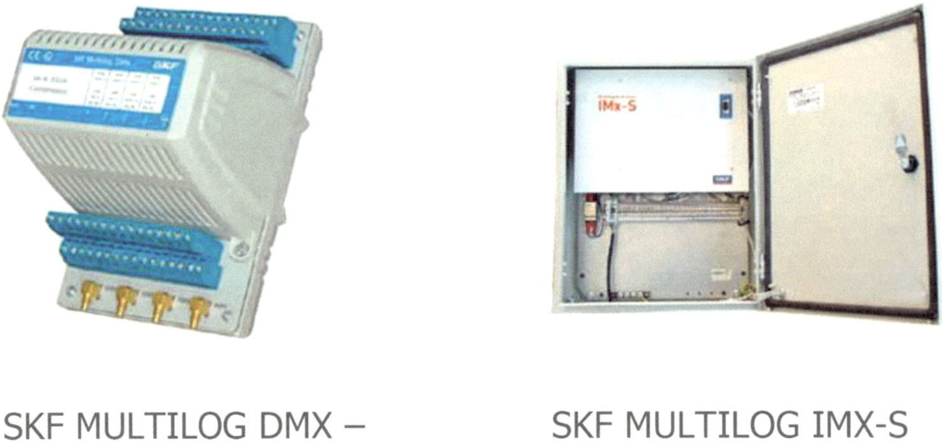

Figure 10.
SKF MULTILOG DMX (left) – SKF MULTILOG IMX-S (right), monitoring & analysers, for monitoring vibration performance [30].

are being designed and manufactured. A common example in the industry is on-line and/or off-line performance monitoring instruments for electrical motors. By way of those instruments it is possible to monitor and detect insulation defects, broken rotor rods, torque problems, load problems and power problems (**Figure 9**) [29].

On the above given figures (**Figure 10**) [31] two instruments are the sample for on-line vibration performance monitoring by, SKF Multilog IMX system is designed for on-line monitoring and it is capable of making simultaneous analysis of a potential failure. SKF Multilog DMX system is designed for protecting the turbo machinery or turbines and is capable of analysis [31].

8. The benefits of predictive maintenance policies

Experience and feedback data from the field indicates that 'Predictive maintenance' provides the following benefits;

1. Maintenance costs of the production department is decreased.

2. Equipment life time capacity is increased.

3. More precise detectability of falures in advance of occurrence.

4. Higher gains in ROI (Return of Investment).

5. The equipment breakdown durations is decreased.

6. Production quality, sustainability and outputs is increased due to pro-active and advance intervention into maintenance process.

7. Increases safety.

8. Reduces interruptions of services.

9. Greater customer satisfaction.

However in order to obtain above defined benefits, advanced digital infra-structure must be established and a platform of IoT (Internet of Things) must be placed. Futhermore in order to analyze the machinery data, machine learning and forecasting based modeling statistical techniques must be utilized [12].

8.1 Advanced maintenance implementations

Advanced Maintenance Implementations are also divided into two groups;

- 1.c.1 Reliability Centered.

- 1.c.2 Risk Based Maintenance.

8.2 Reliability centered maintenance

(RCM)-As a part of advanced maintenance technique, 'reliability-centered maintenance (RCM)' is the optimized mix of reactive, time or interval-based, condition-based, and proactive maintenance types.

The objective of RCM, to generate optimized maintenance plans. Since the maintenance is defined as the sum of technical and administrative activities in order to protect the equipment integrity and it is made for enabling equipment to fulfill its intended functions. On the other hand RCM is the maintenance activities that are implemented for an equipment to fulfill its functions in a manner that is technically in compliance, feasible and approved economically.

Those defined duties; must protect equipment functions, prevent the unexpected premature breakdowns and mitigate the effects of those failures when they happened. Overall objective of a RCM is to intersect the plant reliability and profitability with pro-active maintenance amount on an optimum point. In this way with RCM an optimized maintenance plan is generated for the concerned equipment. RCM is an indicator to show the compliance with company policies and standards. RCM adjusts the maintenance levels and required resources.

In the RCM approach, there are seven basic questions that are implemented with the main lines;

- What kind of functions and performance standards must be fullfilled for a physical entity in the plant existing operation conditions?
- What kind of obstacles are there to prevent fulfilling those functions?
- What are the basic causes for functional failures?
- What are the consequences if a failure is realized?
- How the failure is happened?
- What actions could be taken for preventing and detecting failures before it happens?
- What have to be done if a pro-active maintenance method is not available?

RCM analysis are made with a team that is selected from the representatives of Operations, Process technologies, Maintenance group, Special engineering units (material, equipment, etc) Basic steps of RCM could be defined as; 1. Equipment information, 2. Prominent failure modes, 3. Failure scenario and criticality, 4. Failure modes features & duties, 5. Economic verification, 6. Grouping maintenance duties and implementation, 7. Analysis, feedback &review 8. Equipment Selection.

For RCM analysis, by considering criticality equipment selection is made. The criticality in here is lack of maintenance for an equipment or maintenance in case of failure and the associated risk for this situation. Risk analysis is made with evaluating the failure results and probability; Health and Working Safety (Fire, toxic wastes, and gases), Economic losses (production loss, maintenance cost), Environmental (Leakages, un-controlled emissions,..), With the criticality evaluation here, Equipment list will be defined for RCM analysis.

Equipment Information: In order to detect the failure mechanism (failure mode) and realizing equipment information is needed. Failure mode consists of a model for defining the failure type for a part of equipment. A failure mode is defined as such;

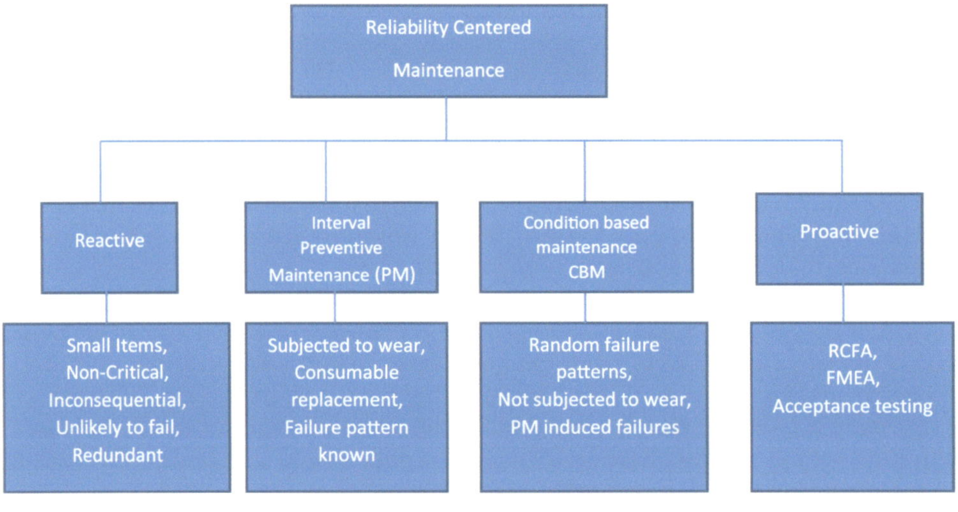

Figure 11.
Components of reliability centered maintenance (RCM) program [33].

Object + Failure Definition Prominent failure modes: Prominent failure modes must contain all failures that could be progressed in systematic ways. Frequency of these failures must be 20 years or more frequent. In listing the Prominent failure modes, investigating operational conditions of equipment and evaluation of local conditions are important along with the knowledge and experience. In here Pareto approach is utilized for distinguishing the most important few failures from less important many failures. In reality most of the reasons of breakdown one or two failure mode [32].

Reliability-centered maintenance components: The components of RCM program are shown in **Figure 11**, [33]. This figure showing that RCM program consists of (reactive maintenance, preventive maintenance, condition based maintenance, and proactive maintenance (System Root Cause Failure Analysis (RCFA), Failure mode effect analysis (FMEA), Acceptance testing).

Basic steps of RCM are defined as follows:

- Step1: selection of system and data collection.
- Step2: definition of system boundary.
- Step3: description of system and functional block.
- Step4: system function functional failures.
- Step5: implementation of failure mode effect analysis.
- Step6: logic tree diagram.
- Step7: task selection [34].

Determining the list of the basic system components is one of the first stage in definition of RCM. The criticality analysis requires different kind of data of each component that build up the system. The effect of failure of the system main components

Criteria	Weight	Levels
Impact on production P	30%	(3) Very important
		(2) Important
		(1) Normal
Impact on safety S	30%	(3)Very important
		(2) Important
		(1) Normal
Availability of standby A	25%	(3) Without standby
		(2) With stand by and medium availability, and
		(1) With standby and high availability
Equipment value V	15%	(3)High value
		(2) Normal, and
		(1) Low value

Table 3.
Criticality analysis in Reliability Centered Maintenance (RCM) [35].

may effect system productivity and maintenance cost. The factors effecting selection of critical system are as follows:

1. Mean-time between failures (MTBF).

2. Total maintenance cost.

3. Mean time to repair (MTTR).

4. Availability.

In the implementation of RCM, some of the well known reliability analysis methods are utilized such as Logic Tree Analysis (LTA), Failure Mode Effect Analysis (FMEA), Failure Mode Effect Criticality Analysis (FMECA).

In usual applications in RCM, in order to perform failure modes, effects and criticality analysis (FMEA/FMECA) the identification of the following basic information has to be defined as indicated in **Table 3** [35].

9. Risk based maintenance

The importance of maintenance function is becoming increased in the recent days in various industrial sectors. Risk Based maintenance (RBM) is a maintenance policy and strategy that combines the advantages of traditional maintenance methods with Risk Based Inspection (RBI) methods and focuses on the mechanical integrity context. RBM helps to select the most effective maintenance strategy depending on the equipment state score and reliability parameters for the whole equipment in the plant. The construction of RBM necessitates the recording of equipment failure and maintenance history in order to define the existing condition of equipments. The data that is collected via RBI, while added into equipment or system history will provide technical inputs for risk analysis. Natural result of this process is to generate the control & maintenance plans for each equipment in order to achieve a reliable and safe plant.

Risk Based Inspection, sheds the lights on the areas of having mechanical integrity that defines the risks that are not defined with other organizational risk analysis methods. In this context RBI becomes a tool for risk analysis and risk management.

To integrate the RBI studies into corporate risk management activities, will be the key factor for the success of risk management program.

To maintain the equipments maintenance and safety in the plant with economic and technical competency, and to consider and manage the problems such as corrosion, erosion, operational and environmental impacts, that might be the causes of breakdowns and stoppages, reducing maintenance costs and down times are the issues that are not fully focused yet by the maintenance sector [36].

Decision-making is required in establishing an optimum maintenance plan, and RBM can play a significant contribution in this stage. But even before RBM was introduced, experienced old personnel in the company had probably made their own decisions on items such as how to obtain optimum result, what inspections are more suitable and what kind of parts/components needs to be prioritized to maintain safety. Presence of such kind of highly skilled old craftsmen could lead to the high level of reliability for devices and or equipment [37] (**Figure 12**).

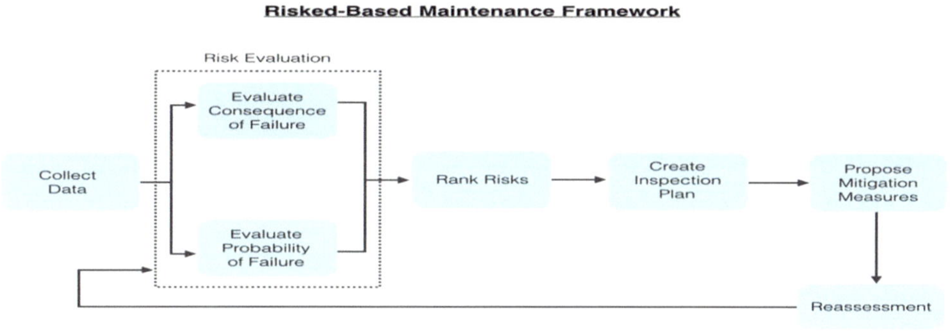

Figure 12.
Risk based maintenance framework [38].

Risk-based maintenance schemes favors low maintenance costs while maintaining a high reliability of the grid, because maintenance measures/actions are planned according to the technical condition of the equipment and the consequences in case of failures only. An investigation in this regard of the individual loss of energy due to failures at the ring main unit of Medium Voltage (MV)/Low Voltage (LV) substations is selected to evaluate the importance of the characteristics of the grid and the station [39].

10. Corrective maintenance

10.1 Corrective (Reactive) maintenance (CM)

Corrective maintenance is also called as reactive maintenance. Corrective maintenance is realized upon observing or detecting a breakdown. In most cases Corrective Maintenance is made after the equipment-machinery breakdown-failure or detecting any equipment problem.

Corrective maintenance is usually encountered in the companies that planned maintenance is not regularly adopted or embraced.

Some examples for corrective Maintenance (CM):

- During the normal production flow a sudden breakdown may occur, the equipment is stopped, and urgent intervention may be needed. Such maintenance is called as corrective-reactive maintenance.

- An important wear that could cause further failure could be detected during a routine equipment inspection, in that case a corrective maintenance is realized in order to prevent further damages and production delays the worn out part is replaced.

- In some cases a simple and cheap equipment part is detected as faulty, but replacement is not made until the part is broken down completely.

In most cases for the corrective maintenance (CM) implementation cases usual score for Mean Time to Repair (MTTR) is realized in longer duration than the expection. Furthermore.

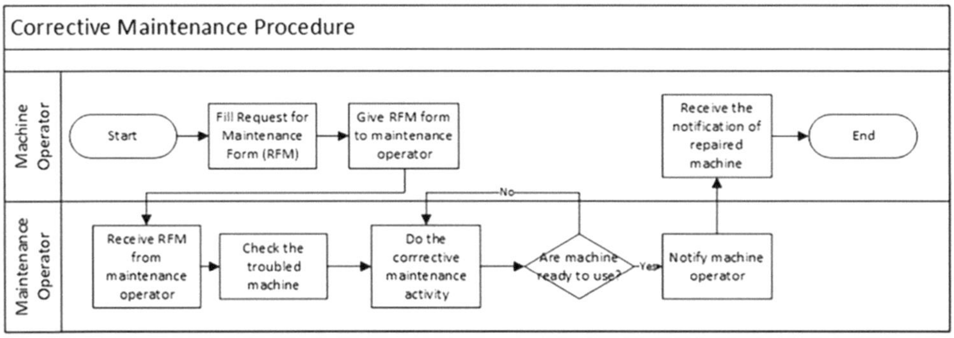

Figure 13.
Corrective maintenance procedure work flow [41].

During the CM processes root causes for failures are not dealt with hence mean time between failure (MTBF) parameter could result in lower durations than expection. As a result within a short period of maintenance sequence a lot of repeated failures are encountered [40].

Corrective Maintenance Procedure Work Flow is given with (**Figure 13**) [41].

10.2 Types of corrective maintenance

Corrective maintenance may be classified under the following categories.

Corrective repair: This kind of equipment repair is made after detecting/observing the failure in order to recover the problem to normal functioning state.

Its operational state.

Basic overhaul: This kind of repair is made to restore the equipment overall parts to their normal functioning state for over-burdened equipment regardless of detecting any specific failure.

11. Method

Salvage: This kind of repair is made usually for worn out aged equipment that is not feasible using after reapair, usually corrective maintenance for that kind equipment is made for selling the equipment with reasonable price.

Servicing: This kind of corrective maintenance may require external expert supplier maintenance intervention such as engine cylinder & piston repairing or replacement.

Rebuild: This is a rather costly CM maintenance operation. If a critical equipment/machine can not be replaced easily in that case rebuilding complete parts and body ciuld be inevitable. In that case by considering original equipment maintenance/service manual re-building the whole parts is attempted regardless of the higher maintenance costs [42].

11.1 Improvement strategies in corrective maintenance effectiveness

In order to improve corrective maintenance performance, corrective maintenance duration has to be reduced. Some of the useful measures to improve corrective maintenance effectiveness is given below:

- Proper design has to be made in order to reach the equipment components easily,
- As less parts as possible to be dismantled to reach the repair location,
- There must be sufficient room to enable operator working properly in the maintenance operator working space,
- Vision convenience has to be provided during corrective maintenance,
- Standard and/or interchangeable parts has to be preferred in the equipment body to enable demounting with various tools, and to reduce corrective maintenance duration,
- Hole lids are to be provided with openable as minimum 180° or to be complete demountable,
- Ergonomic working height to be preferred during corrective maintenance,
- Proper lubrication holes to be provided to enable easy maintenance,
- Detailed proper maintenance work instructions to be provided to detect the faults and failures easily and to react for repair properly [1, 11, 43].

Maintenance parameters	Preventive maintenance	Predictive maintenance
Maintenance cost	HIGHER – planned interval must be regularly implemented	LOWER-maintenance is implemented just before the breakdown occurs or when needed. US Department of Energy research indicated that predictive maintenance is extremely cost-effective. 25–30% reduction of maintenance costs
Failure detection ability	LOWER- Failure detection can be made during the regular maintenance time	HIGHER-Since high technology equipment and/or sensors are utilized, detection will be mostly precise and in earlier times
Return on investment financial gains	LOWER-	HIGHER-A US Department of Energy research indicated that predictive maintenance is extremely cost-effective. Implementing a predictive maintenance software can deliver notable financial gains with a significant ROI
Number of break-downs	HIGHER	LOWER-US Department of Energy research indicated that 70–75% fewer breakdowns
Reduction of downtime	LOWER	HIGHER-US Department of Energy research indicated that 35–45% downtime decline
Infra structure cost	LOWER	HIGHER-Set up of infrastructure, relevant hardware and software, provision of sensors and training relevant operators will be extra costly [44]

Source: [44]

Table 4.
Comparison of preventive and predictive maintenance policies [44].

At the end of the chapter preventive and predictive maintenance activities are compared with **Table 4** [44].

12. Results and conclusions

In this chapter basic definitions are made for preventive (planned) maintenance, predictive (condition based) maintenance and assessment of policies for those maintenance types.

Additionally advanced maintenance types (with reliability and risk based maintenance) and reactive (corrective) maintenance types are also defined. And comparison of preventive and predictive maintenance is made in terms of the Maintenance Cost, Failure Detection Ability, Return on Investment Financial Gains, Number of breakdowns, Reduction of downtime, Infra structure cost,....etc.

It is concluded that in most of the parameters, predictive maintenance have superior features over the other maintenance policies.

We could argue that with regard to Total Productive Maintenance (TPM) approach predictive maintenance policy is the most effective type. Followed by preventive maintenance and advanced maintenance (reliability based and risk based) policies. Companies willing to adopt Total Quality Management approach should switch from reactive (corrective) maintenance into preventive (planned) and predictive (condition based) maintenance policies.

Author details

Hikmet Erbiyik
Faculty of Engineering, Department of Industrial Engineering, Yalova University, Yalova, Turkey

*Address all correspondence to: hikmeterbiyik53@gmail.com

References

[1] Available from: https://www.managementstudyguide.com/maintenance-policy-and-repair.htm

[2] Dejan B, Zdravko M. The Role and Importance of Planning of Maintenance in Industrial Practice. Prijevor, Serbia: DQM International Conference Life Cycle Engineering and Management; 2019

[3] Duc T, Karol D, Katarzyna S. The predictive maintenance concept in the maintenance department of the "Industry 4.0". Production Enterprise Foundations of Management. 2018. ISSN 2080-7279;**10**. DOI: 10.2478/fman-2018-0022

[4] Ötleş S, Çolak UC, Ötleş O. Artificial intelligence for industry. Plastic Packaging Journal (PlastikAmbalajDergisi). 2018:46-50

[5] Girdhar P, Girdhar and Associates. In: Scheffer C, editor. Practical Machinery Vibration Analysis and Predictive Maintenance. IDC Technologies; 2004;**1**

[6] Amran TG, Sujarto L. Early warning system in preventive maintenance as a solution to reduce maintenance cost. In: Proceedings of the 2014 International Conference on Industrial Engineering and Operations Management. Bali, Indonesia; 2014. pp. 596-605

[7] Lacey SJ. The Role of Vibration Monitoring in Predictive Maintenance Engineering Manager Schaeffler (UK) Limited, Available from: https://www.schaeffler.com. Vol. 202022. p. 03

[8] IRCM, Institute of reliability centred maintenance. 2022

[9] Kurgan N. Production Management and Organization. Vol. 19. Mak 404: Mayis University; 2021

[10] Available from: https://whatispiping.com/types-of-maintenance/. 2022

[11] Available from: https://www.upkeep.com/learning/planned-maintenance

[12] Muhammad S. Lecturer, University of Gujrat, Pakistan, AFA Workshop on "Quality Control & Assurance in Maintenance in Fertilizer Industries". Oman; 2015. p. 23

[13] Gürsoy VD. International Journal of 3d Printing Technologies and Digital Industry 3:1. Vol. 56-662019. p. 61

[14] Power Engineering International, Timothy P. Holtan, Smart Signal Corporation, Illinois, USA, 2003

[15] Available from: https://www.sensegrow.com. 2022

[16] Understanding the Benefits of Vibration Monitoring and Analysis. 2022. Available from: https://www.fluke.com

[17] Emil C, Lucian P, Petrescu M-C. The major predictive maintenance actions of the electric equipments in the industrial facilities. 2017;**1**:26-33. DOI: 10.1515/SBEEF-2017-0018

[18] Venegas P, Ivorra E, Ortega M, Márquez G, Martínez J, Sáez de Ocáriz I. Development of thermographic module for predictive maintenance system of industrial equipment. Porto: 15th Quantitative InfraRed Thermography Conference; 2020. pp. 6-10

[19] Güngör A, Cemal M. Failure detection and prevention in cement production sector via predictive maintenance. Engineer and Machinery. 2018;**59**(692):48-67

[20] Mobley RK. An Introduction to Predictive Maintenance. New York: Van Nostrand Reinhold; 1990. pp. 1-16

[21] Sarah K, Terence LVZ. Automating Predictive Maintenance Using Oil Analysis and Machine Learning. 2020. DOI: 10.1109/SAUPEC/RobMech/PRASA48453.2020.9041003. Conference: 2020 International SAUPEC/RobMech/PRASA Conference

[22] Zhu J, Yoon JM, He D, Qu Y, Bechhoefer E. Lubrication oil condition monitoring and remaining useful life prediction with particle filtering. International Journal of Prognostics and Health Management. 2013;**4**:124-138

[23] Sharma B, Gandhi O. Performance evaluation and analysis of lubricating oil using parameter profile approach. Industrial Lubrication and Tribology. 2008;**60**(3):131-137

[24] Gülşen Y, Murat KH. Failure detection with vibration analysis in pumps and a case study for predictive maintenance. Journal of Installment Engineering. 2014;**140**

[25] Available from: http://megep.meb.gov.tr/mte_program_modul/moduller_pdf/Korozyon%20Ve%20Katodik%20Koruma.pdf

[26] Homborg AM, Tinga T, Mol JMC. Listening to corrosion. In: NATO Science &Technology Organization AVT-305 Research Specialists' Meeting on Sensing Systems for Integrated Vehicle Health Management for Military Vehicles. Athens, Greece: Hellenic Armed Forces Officers Club; 2018

[27] Cottis RA. Techniques for Corrosion Monitoring. Second ed. Woodhead Publishing Series in Metals and Surface Engineering; LLC Carson City, NV United States. 2021. pp. 99-122

[28] Allahar KN, Upadhyay V, Bierwagen GP, Gelling VJ. Monitoring of a military vehicle coating under Prohesion exposure by embedded sensors. Progress in Organic Coating. 2009;**65**:142-151

[29] SKF – Baker AWA IV – Offline Test Instrument SKF – Baker The Explorer – Online Test Instrument. 2022. Available from: https://silo.tips/queue/bakmn-sisteminin-nemi?&queue_id=-1&v=1651596391&u=OTUuNzAuMjA3LjE4

[30] SKFMULTILOGDMX–SKFMULTILOGIMX-S, Monitoring & Analysers. 2022. Available from: https://silo.tips/queue/bakmn-sisteminin-nemi?&queue_id=-1&v=1651596391&u=OTUuNzAuMjA3LjE4

[31] Available from: https://silo.tips/queue/bakmn-sisteminin-nemi?&queue_id=-1&v=1651596391&u=OTUuNzAuMjA3LjE4. 2022 2017;1:1-61

[32] Available from: https://www.emo.org.tr/ekler/b7cab6e498c4424_ek.pdf

[33] Alan Pride CMRP. Associate director systems reliability, reliability-centered maintenance (RCM). Smithsonian Institution. 2016

[34] Rausand M. Reliability-centered maintenance. Reliability Engineering and System Safety. 1998;**60**(2):121-132

[35] Afefy IH. Reliability-centered maintenance methodology and application: A case study. Engineering. 2010;**2**:863-873. DOI: 10.4236/eng.2010.211109. Published Online November 2010 (http://www.scirp.org/journal/eng)

[36] Available from: https://www.adakalite.com/hizmetlerimiz/risk_tabanli_bakim/

[37] Sakai S. Risk based maintenance. JR EAST Technical Review. 2010;**17**:1-4

[38] Risk Based Maintenance Framework. 2022. Available

from: https://www.fiixsoftware.com/maintenance-strategies/risk-based-maintenance

[39] Köhn P, Schnettler A, Schultze N. Analysis of condition and risk-based maintenance planning for medium voltage/low-voltage substations. In: 24th International Conference & Exhibition on Electricity Distribution (CIRED). Glasgow-UK: IET Journals, The Institution of Engineering & Technology; 12-15 Jun 2017. DOI: 10.1049/oap-cired.2017.0774. Available from: www.ietdl.org

[40] Available from: https://www.repairist.com.tr/duzeltici-bakim-nedir/

[41] Available from: https://leanmanufacturing.online/support-autonomous-maintenance/planned-corrective-maintenance-flow-chart/ 2022

[42] Dhillon BS. Engineering Maintenance: A Modern Approach. Boca Raton, FL: CRC Press; 2002

[43] Babacan UC, Meran C. Design practices toward maintenance. Engineer and Machinery. 2019;**60**(695):119-131, Review Article

[44] Available from: https://fieldcode.com/en/resources/blog/5-advantages-of-predictive-maintenance-and-how-to-leverage-on-them

Chapter 3

Maintenance and Renewal Cost Evaluation for Managing Assets of Electric Power Equipment and Operational Data Analysis for Failure Rate Estimation

Tsuguhiro Takahashi

Abstract

In recent years, "asset management" or "managing assets" technique has been expected to rationalize maintenance and operation of electric power equipment, especially for aging equipment. Some concrete support tools have been developed by considering life cycle cost for substation equipment in "Central Research Institute of Electric Power Industry, Japan," which include failure risk evaluation. Such cost and risk evaluation are essential for comparative evaluation of different types of equipment. Failure probability is one of the most important factors for the evaluation. Because of its high reliability, electric power equipment can be expected to have a very long lifetime, therefore, durability test is not applicable, but rather relies on analysis of actual operational data. Collection, accumulation, and analysis of actual operational data are necessary for accurate evaluation. This chapter describes the evaluation method for the managing assets, and data collection and analysis to improve the accuracy of failure probability estimation.

Keywords: asset management, power equipment, preventive maintenance, life cycle cost, risk evaluation

1. Introduction

In general, power transmission and distribution equipment that can be expected to operate for more than several decades had a wide age distribution over time. The amount of capital investment is inevitably affected by society and the economy, therefore, the shift to the aging side of the age distribution is progressing as a common phenomenon in many countries in recent years. In US and European countries, the issue of aging has been discussed since the end of the 20th century [1], and the effective introduction of the so-called asset management technique for formulating maintenance and management strategies that appropriately balance risk and cost has been examined.

2. Maintenance and renewal cost evaluation for power transmission and distribution equipment

In the asset management for corporate activities, benefits as positive impacts and risks as negative impacts are generally evaluated and added for each possible activity to be considered as an evaluation index to select the optimal strategy. In the case of electric power transmission and distribution equipment, it is difficult to evaluate the contribution of individual equipment because the entire network system generates benefits, therefore in many cases, the optimal maintenance strategy is selected by minimizing the costs and statistically evaluating risks required to maintain the network size and reliability. As one simple model, CRIEPI has proposed the cumulative cost evaluation method, and some support programs have been developed [2–5].

2.1 Maintenance and renewal "cost" in operation

Maintenance and renewal costs during normal operation are classified into the following four items based on their expenditure timing and characteristics of change by age.

i. Average repairing cost.

   In ordinary operations, there are some necessary repairing costs, such as oil leakage repair cost for power transformers. Generally, it can be assumed to increase with age. For example, its characteristic is assumed to be proportion to age.

ii. Inspection cost.

   Generally, regulated inspection cost is needed commonly for each equipment. A periodic and a nonperiodic (which is performed at a certain age) inspection costs are considered.

iii. Overhaul cost.

   Some equipment can be applied so-called "Overhaul" to realize rejuvenation as a maintenance measure. Overhaul costs depending on their effect is considered.

iv. Installation cost of planned renewal.

   The installation cost of equipment can be regarded as installments over several years, by considering depreciation. The property tax should also be considered during these years.

2.2 Statistically expected failure cost as "risk"

The expense required at a failure is installation cost of renewed equipment and some so-called penalty costs. The "penalty" cost should include the lost revenue from selling electricity, the emergent recovery cost, a penalty resulting from service interruption, and so on. Since the occurrence of a failure is statistical, the expense is "statistically expected cost," which is the product of the "cost of failure" and the "probability of failure." This cost is not a real cash flow, but it is expressed in monetary values and can be compared and combined with maintenance and renewal costs. When some aged equipment in service is removed as a result of failure, the same number of new equipment should be installed in order to maintain its power network scale. That is, the total number of equipment does not change. From the statistical point of view, such failures are occurred every year, depending on their failure probability. This means that the age distribution changes over time, which should be considered when the cumulative cost evaluation is carried out [4].

2.3 Maintenance scenarios to be compared

In general, when the asset management techniques are utilized for maintenance and renewal planning, it is necessary to consider possible maintenance measures and scenarios, in advance. The cumulative cost evaluation should be carried out for each scenario. Therefore, this scenario setting is important for this method. As one example, time-based renewal scenarios (such as at 40 and 50 years) with and without overhaul (OH) have been considered, as shown in **Figure 1**. The OH is assumed to rejuvenate equipment at a certain cost. Its effect (rejuvenation years), cost, and timing are specified as parameters.

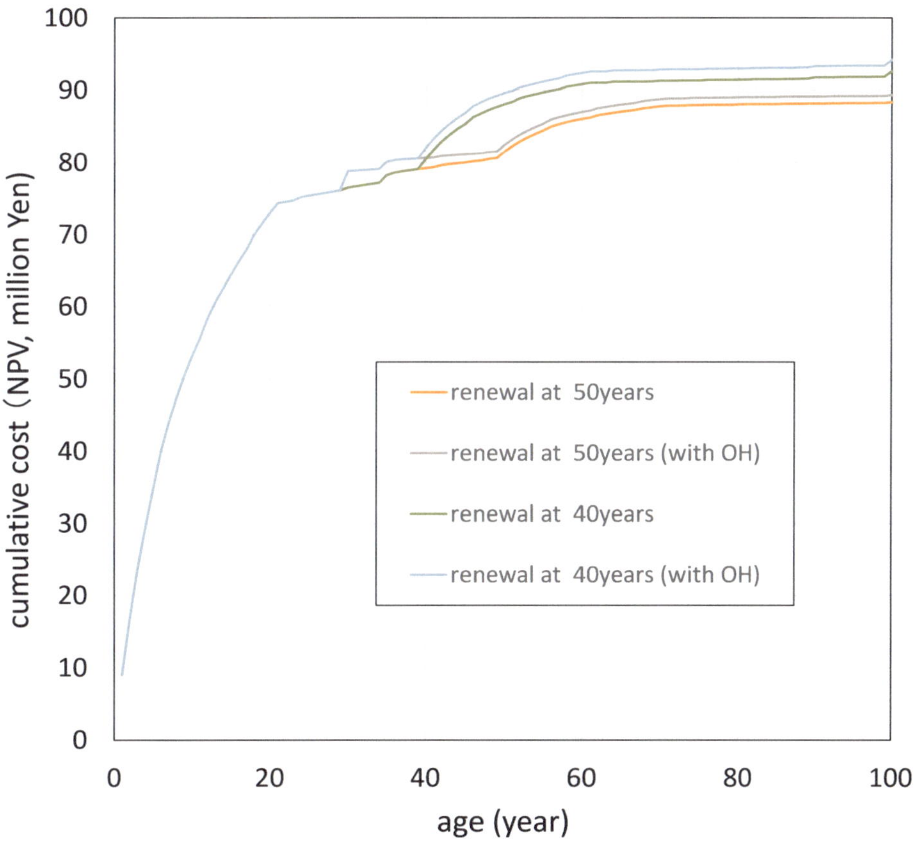

Figure 1.
Output example of CRIEPI's support program.

3. Failure data analysis based on operation results

One of the most important items in the managing assets is the risk evaluation, and failure probability estimation for each equipment is crucial. The failure probability distribution of a product is generally obtained from endurance tests on a large number of the same products. However, it is not practical to conduct such endurance tests for electric power transmission and distribution equipment, which can generally be expected to operate for more than several decades. In order to investigate the failure probability characteristics of long-life products, statistical analysis of residual performance tests of removed products from a real field, and operation/failure results in a real field are often employed. This section describes statistical analysis methods for operational data.

3.1 Definition of failure rate

Failures that are generally considered with a failure rate for an industrial product include those that stop the operation of the product once they occur and those that are repaired repeatedly each time they occur. The former determines the service life of the product, and the rate of occurrence is usually evaluated as a function of operating hours. The latter usually focuses on the interval of occurrence of multiple failures, and the change over time of the average time or the occurrence rate in a certain operating time is evaluated. In this section, "failure rate" means the occurrence rate of the former failures at a certain age, and is expressed as a function of age.

If a large number of the same products start operating simultaneously, the percentage of those that continue to operate without failure up to a certain elapsed time t is generally called the reliability, and it is often expressed as $\mathrm{R(t)}$. The cumulative failure probability, which means the percentage of failures between the start of operation and t, is expressed as follows:

$$\mathrm{F(t)} = 1 - \mathrm{R(t)} \tag{1}$$

$\mathrm{F(t)}$ differentiated by t is often denoted as $\mathrm{f(t)}$.

$$\mathrm{f(t)} = \frac{d}{dt}\mathrm{F(t)} \tag{2}$$

This is the increment of $\mathrm{F(t)}$ at time t, i.e., the percentage of products that fail at time t, for all products. It is sometimes called the "failure probability" because it is the time derivative of the "cumulative failure probability," but it is also called the probability density of failure because it represents the probability distribution of when the product will fail. When examining the risk of equipment in service, the probability that equipment that has been operating until age t will fail by the following year is often utilized. This is often denoted as $\lambda(t)$ and is obtained as follows:

$$\lambda(t) = \frac{\mathrm{f(t)}}{\mathrm{R(t)}} = \frac{d\mathrm{F(t)}}{dt} \bullet \frac{1}{\mathrm{R(t)}} = -\frac{d\mathrm{R(t)}}{dt} \bullet \frac{1}{\mathrm{R(t)}} \tag{3}$$

In this chapter, this is referred to as the "failure rate," and a method for estimating it from operation results is discussed.

3.2 Characteristics of power transmission and distribution equipment operational data for failure rate estimation

In general, to examine the failure rate characteristics of a product, an endurance test is conducted using several units of the same product. To examine aging characteristics, test samples are usually operated simultaneously and the time required for each sample to reach failure is determined. In conducting endurance tests, it is not always possible to continue the test until all samples fail due to time and cost constraints, but some statistical analysis methods can analyze data obtained by discontinuing the endurance test in the middle of the test. In the case of power transmission and distribution equipment, it is difficult to plan an endurance test in which a sufficient number of test samples are operated simultaneously, but a method to estimate the failure rate by considering the actual results of the long-term operation of a large number of facilities in a real field as a pseudo endurance test is conceivable. In this case, some considerations need to be made for the data used in the analysis.

3.2.1 Data for failure rate estimation

In order to estimate the failure rate, information on equipment that has been in operation without failure is needed in addition to the equipment that has failed. For a group of equipment that is the same type and assumed to exhibit the same failure rate aging characteristics, it is necessary to investigate the age of each failed facility as well as the age distribution of the group in operation.

3.2.2 Observation period

Failure rate estimation based on operational data for a group of equipment with age distribution starts by determining the number of failures/operation equipment at each age in order to obtain an approximation of $\lambda(t)$. In doing so, it is necessary to consider the period of the survey (observation period). For example, if the past 10 years of failure history is to be investigated, the number of failures for each age can be obtained by adding up the 10 years of information for each individual failure by age of occurrence. Similarly, for the number of operations, the actual number of operations (the age distribution of equipment in operation) for each year of the observation period should be surveyed and added. For example, equipment that has continued to operate without failure at 20 years old has existed every year for the past 10 years, so they can be added up to the amount of equipment in operation at 20 years old in the aggregate. However, for example, equipment that is currently 40 years old should not be combined as "equipment that has continued to operate at 20 years old without failure" because this information is outside the observation period, though it is clear that the equipment continued to operate for 20 years without failure 20 years ago. This is because the information on equipment that has failed or has been removed outside the observation period cannot be used. Only "should have been in operation" information is likely to lead to underestimation of the failure rate. Therefore, the "observation period" is important and should be paid attention in the failure rate estimation.

3.2.3 Influence of low number of failure results

The inherent difficulty in statistical analysis of the failure rate from operational data of power transmission and distribution equipment lies in the fact that such equipment has a low failure rate and is highly reliable, and that preventive maintenance is performed to maintain high supply reliability. These suppress the occurrence of failures during operation, resulting in a decrease in the accuracy of failure rate estimation and underestimation. There is no other way to deal with these problems than to continuously accumulate appropriate data and increase the amount of data.

3.3 Procedure of statistical analysis

This section presents a computer simulation of virtual operational data and uses the results to show a specific procedure for estimating failure rates [6].

3.3.1 Simulated data

The simulating equipment is a group of 12,340 units with the aging distribution shown in **Figure 2**, all of which are assumed to have the same failure rate characteristics shown in **Figure 3**. The Weibull distribution is assumed for the failure rate characteristics, and the failure rate $\lambda(t)$ and probability density of failure $f(t)$ are expressed as follows:

$$\lambda(\mathrm{t}) = \frac{m}{t_s} \cdot \left(\frac{t}{t_s}\right)^{m-1} \tag{4}$$

$$\mathrm{f}(\mathrm{t}) = \frac{m}{t_s} \cdot \left(\frac{t}{t_s}\right)^{m-1} \cdot \exp\left\{-\left(\frac{t}{t_s}\right)^{m}\right\} \tag{5}$$

where m is the shape parameter and t_s is the scale parameter, and in **Figure 2**, m and t_s (years) are set to 4 and 80, respectively.

It can be simulated that after 1 year of operation of this equipment group, some equipment will fail according to the failure rate determined by each age. For each aged equipment, a random number between 0 and 1 is generated, and when the value is less than the failure rate at its age the unit is regarded to fail. In the following year, the number of failures is subtracted from the amount of equipment in each age, and the age distribution is shifted to the higher side by 1 year, and new equipment equal to the number of failures in the previous year is regarded to be installed, assuming

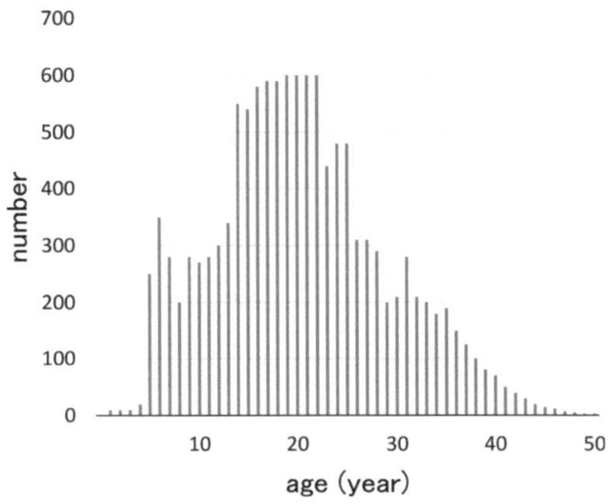

Figure 2.
Assumed age distribution (12,340 units).

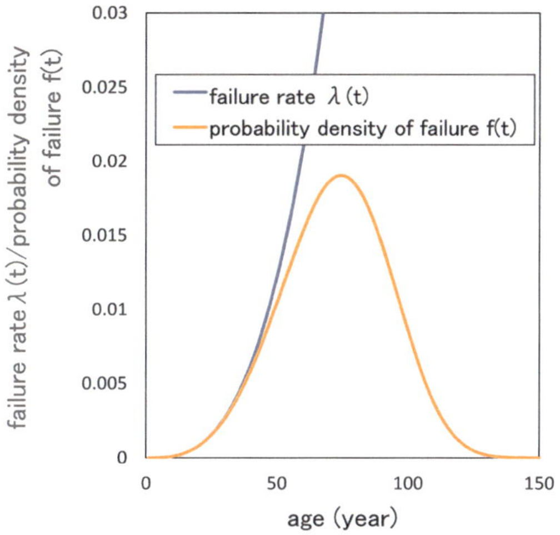

Figure 3.
Assumed failure characteristics.

maintenance that keeps the total amount of equipment in the group constant. If this is continued for several years, which corresponds to "observation period," operational data can be generated as one simulation case, but its result would depend on the random number output, so different results would be obtained each time of the simulation. One simulation result whose observation period is 5 years is shown in **Table 1**. **Table 1** also includes the combined number at each age during the observation period. The statistical analysis in the following section is performed on this combined operational data. Such simulation was performed five times.

3.3.2 Hazard analysis

Failure characteristics of high voltage equipment are often expressed as a Weibull distribution. In other words, the fitting of an endurance test and operational performance data is performed assuming that the probability density distribution of failure can be expressed as a Weibull distribution function. The Hazard analysis is a statistical analysis method for this purpose. In this simulation, as described in Section 3.3.1, the true distribution is given as Weibull distributions, therefore, if hazard analysis is performed with high accuracy, the distribution is expected to be restored.

When the failure characteristic follows a Weibull distribution, the failure rate $\lambda(t)$ is expressed by Eq. (4). This is integrated over time as in the following equation and is called the cumulative hazard $H(t)$.

$$\mathrm{H(t)} = \int_0^t \lambda(\tau)d\tau = \left(\frac{t}{t_s}\right)^m \tag{6}$$

Taking the natural logarithm of both sides of Eq. (6), the following equation is obtained.

$$\ln\{\mathrm{H(t)}\} = \ln\left\{\left(\frac{t}{t_s}\right)^m\right\} = m \bullet \ln(t) - m \bullet \ln(t_s) \tag{7}$$

age (year)	1st year		2nd year		3rd year		4th year		5th year		total	
	operating	failed	operating	failed	operating	failed	operating	failed	operating	failed	operating	failed
1	10	0	22	0	22	0	23	0	25	0	102	0
2	10	0	10	0	22	0	22	0	23	0	87	0
3	10	0	10	0	10	0	22	0	22	0	74	0
4	20	0	10	0	10	0	10	0	22	0	72	0
5	250	0	20	0	10	0	10	0	10	0	300	0
6	350	0	250	0	20	0	10	0	10	0	640	0
7	280	0	350	0	250	0	20	0	10	0	910	0
8	200	0	280	0	350	0	250	0	20	0	1100	0
9	280	0	200	0	280	0	350	0	250	1	1360	1
10	270	0	280	0	200	0	280	0	350	0	1380	0
11	280	0	270	0	280	0	200	0	280	0	1310	0
12	300	1	280	0	270	0	280	0	200	0	1330	1
13	340	0	299	0	280	0	270	0	280	0	1469	0
14	550	0	340	0	299	0	280	0	270	0	1739	0
15	540	0	550	0	340	0	299	0	280	1	2009	1
16	580	0	540	0	550	0	340	2	299	0	2309	2
17	590	0	580	0	540	0	550	0	338	0	2598	0
18	590	1	590	1	580	0	540	0	550	0	2850	2
19	600	1	589	0	589	0	580	0	540	0	2898	1
20	600	2	599	1	589	2	589	0	580	0	2957	5
21	600	0	598	1	598	2	587	1	589	1	2972	5
22	600	0	600	0	597	0	596	1	586	0	2979	1

age (year)	1st year		2nd year		3rd year		4th year		5th year		total	
	operating	failed	operating	failed	operating	failed	operating	failed	operating	failed	operating	failed
23	440	0	600	0	600	2	597		595	1	2832	3
24	480	3	440	0	600	0	598	1	597	4	2715	8
25	480	1	477	1	440	0	600		597	2	2594	4
26	310	0	479	0	476	2	440	1	600	1	2305	4
27	310	1	310	3	479	0	474		439	0	2012	6
28	290	0	309	2	307	0	479	1	472	1	1857	4
29	200	1	290	1	307	1	307	0	478	0	1582	3
30	210	1	199	0	289	0	306	2	307	0	1311	3
31	280	1	209	1	199	0	289	0	304	0	1281	2
32	210	2	279	0	208	0	199	0	289	2	1185	4
33	200	1	208	0	279	3	208	0	199	1	1094	5
34	180	2	199	0	208	1	276	0	208	1	1071	4
35	190	1	178	1	199	4	207	0	276		1050	9
36	150	0	189	1	177	0	195	1	207	1	918	3
37	125	0	150	1	188	0	177	0	194	1	834	2
38	100	1	125	0	149	1	188	1	177	1	739	4
39	80	1	99	1	125	0	148	1	187		639	5
40	70	0	79	2	98	0	125	2	147	1	519	5
41	50	0	70	0	77	0	98	3	123		418	3
42	40	0	50	1	70	2	77	2	95	1	332	6
43	30	0	40	1	49	1	68	1	75		262	3
44	20	1	30	0	39	2	48	0	67	1	204	4

age (year)	1st year		2nd year		3rd year		4th year		5th year		total	
	operating	failed	operating	failed	operating	failed	operating	failed	operating	failed	operating	failed
45	15	0	19	0	30	0	37	0	48	2	149	2
46	12	0	15	0	19	0	30	1	37	0	113	1
47	7	0	12	3	15	0	19	1	29	0	82	4
48	5	0	7	0	9	0	15	0	18	0	54	0
49	3	0	5	0	7	0	9	0	15	0	39	0
50	3	0	3	0	5	0	7	0	9	0	27	0
51	0	0	3	0	3	0	5	0	7	0	18	0
52	0	0	0	0	3	0	3	1	5	0	11	1
53	0	0	0	0	0	0	3	0	2	0	5	0
54	0	0	0	0	0	0	0	0	3	0	3	0
55	0	0	0	0	0	0	0	0	0	0	0	0

Table 1.
One example of simulated operational data.

Plotting as y = $\ln\{H(t)\}$ and x = $\ln(t)$ results in a straight line with slope m and y-intercept $-m\ln(t_s)$. Utilizing Eq. (7), the natural logarithm of the approximate value $\hat{H}(t)$ of the cumulative hazard $H(t)$ obtained from endurance tests or operational data is plotted against the natural logarithm of age t, to estimate the shape parameter m and scale parameter t_s by linear approximation. When the number of operations at age t is $N(t)$ and the number of failures is $n(t)$, the approximate value $\hat{\lambda}(t)$ of the failure rate $\lambda(t)$ is obtained from the following equation.

$$\hat{\lambda}(t) = \frac{n(t)}{N(t)} \tag{8}$$

Using $\hat{\lambda}(t)$, $\hat{H}(t)$ is obtained as follows:

$$\hat{H}(t) = \sum_{\tau=1}^{t} \hat{\lambda}(\tau) = \hat{H}(t-1) + \hat{\lambda}(t) \tag{9}$$

Table 2 summarizes the results of these calculations using the combined data in **Table 1**. **Table 2** also includes the natural logarithm of t and $\hat{H}(t)$ for graph plotting. In the hazard plot that shows the relationship between $\ln(t)$ and $\ln\left(\hat{H}(t)\right)$, plotting is carried out only when $\hat{H}(t)$ is changed, that is, when $\hat{\lambda}(t) \neq 0$. The hazard plots created from $\ln(t)$ and $\hat{H}(t)$ in **Table 2** are shown in **Figure 4**. **Figure 4** includes linear approximation, from which the shape parameter m = 3.57 and scale parameter t_s = 82.2 years are obtained from the slope and the intercept.

The Weibull distribution obtained by such procedure is a "sample mean" obtained from the average characteristics of a "sample" of equipment operational data and is expected to be different each time the sample is taken, in this case, each time the operational data is simulated. On the other hand, the true failure rate characteristic, the "population mean," is determined first in this discussion, therefore they can be compared. The failure rates and probability density distributions of failures obtained from the results of five simulations, including the data in **Table 1**, are shown in **Figure 5**. **Figure 5** also shows the true failure rate characteristics. Some of the five estimates (sample mean) have failure rates that are close to the true values, while others are higher or lower. Only one of the sample means, which is expected to vary, can be observed in reality, and there is no way to know the deviation from the true value. The only fundamental solution to the low estimation accuracy due to the low failure rate described in Section 3.2.3 is to increase the amount of data by investigating and accumulating the actual operation of equipment over a long period of time.

3.4 Failure data and preventive renewal data

As mentioned in section 3.2.3, another issue in estimating failure rates from actual operational results is that failures do not occur as actual results because preventive renewals are performed in actual maintenance. This section discusses the addition of renewal data to the failure rate estimation.

Reference [7] introduces the failure rate and renewal rate of transformers and points out that the failure rate does not increase over time but the renewal rate does, and that the failure rate characteristic should be what is shifted renewal rate characteristic to the right (toward the high aging side) if no preventive maintenance is

age t (year)	total		approx. Failure rate $\hat{\lambda}(t)$	approx. Cum. hazard $\hat{H}(t)$	ln (t)	ln ($\hat{H}(t)$)
	operating	failed				
1	102	0	0/102 = 0.0000	0.0000	—	—
2	87	0	0/87 = 0.0000	0.0000	—	—
3	74	0	0/74 = 0.0000	0.0000	—	—
4	72	0	0/72 = 0.0000	0.0000	—	—
5	300	0	0/300 = 0.0000	0.0000	—	—
6	640	0	0/640 = 0.0000	0.0000	—	—
7	910	0	0/910 = 0.0000	0.0000	—	—
8	1100	0	0/1100 = 0.0000	0.0000	—	—
9	1360	1	1/1360 = 0.0007	0.0007	2.197225	−7.21524
10	1380	0	0/1380 = 0.0000	0.0007	—	—
11	1310	0	0/1310 = 0.0000	0.0007	—	—
12	1330	1	1/1330 = 0.0008	0.0015	2.484907	−6.51088
13	1469	0	0/1469 = 0.0000	0.0015	—	—
14	1739	0	0/1739 = 0.0000	0.0015	—	—
15	2009	1	1/2009 = 0.0005	0.0020	2.70805	−6.22217
16	2309	2	2/2309 = 0.0009	0.0029	2.772589	−5.86005
17	2598	0	0/2598 = 0.0000	0.0029	—	—
18	2850	2	2/2850 = 0.0007	0.0036	2.890372	−5.64
19	2898	1	1/2898 = 0.0003	0.0039	2.944439	−5.54731
20	2957	5	5/2957 = 0.0017	0.0056	2.995732	−5.18698
21	2972	5	5/2972 = 0.0017	0.0073	3.044522	−4.92383
22	2979	1	1/2979 = 0.0003	0.0076	3.091042	−48,787
23	2832	3	3/2832 = 0.0011	0.0087	3.135494	−4.74832
24	2715	8	8/2715 = 0.0029	0.0116	3.178054	−4.45565
25	2594	4	4/2594 = 0.0015	0.0132	3.218876	−4.33097
26	2305	4	4/2305 = 0.0017	0.0149	3.258097	−4.20705
27	2012	6	6/2012 = 0.0030	0.0179	3.295837	−4.0245
28	1857	4	4/1857 = 0.0022	0.0200	3.332205	−3.91071
29	1582	3	3/1582 = 0.0019	0.0219	3.367296	−3.82024
30	1311	3	3/1311 = 0.0023	0.0242	3.401197	−3.72095
31	1281	2	2/1281 = 0.0016	0.0258	3.433987	−3.65846
32	1185	4	4/1185 = 0.0034	0.0291	3.465736	−3.53538
33	1094	5	5/1094 = 0.0046	0.0337	3.496508	−3.38972
34	1071	4	4/1071 = 0.0037	0.0375	3.526361	−3.28467
35	1050	9	9/1050 = 0.0086	0.0460	3.555348	−3.07858
36	918	3	3/918 = 0.0033	0.0493	3.583519	−3.00999
37	834	2	2/834 = 0.0024	0.0517	3.610918	−2.96248

age t (year)	total		approx. Failure rate $\hat{\lambda}(t)$	approx. Cum. hazard $\hat{H}(t)$	ln (t)	ln ($\hat{H}(t)$)
	operating	failed				
38	739	4	4/739 = 0.0054	0.0571	3.637586	−2.8629
39	639	5	5/639 = 0.0078	0.0649	3.663562	−2.73448
40	519	5	5/519 = 0.0096	0.0746	3.688879	−2.59613
41	418	3	3/418 = 0.0072	0.0817	3.713572	−2.50423
42	332	6	6/332 = 0.0181	0.0998	3.73767	−2.30448
43	262	3	3/262 = 0.0115	0.1113	3.7612	−2.19587
44	204	4	4/204 = 0.0196	0.1309	3.78419	−2.03356
45	149	2	2/149 = 0.0134	0.1443	3.806662	−1.93591
46	113	1	1/113 = 0.0088	0.1531	3.828641	−1.87639
47	82	4	4/82 = 0.0488	0.2019	3.850148	−1.59987
48	54	0	0/54 = 0.0000	0.2019	—	—
49	39	0	0/39 = 0.0000	0.2019	—	—
50	27	0	0/27 = 0.0000	0.2019	—	—
51	18	0	0/18 = 0.0000	0.2019	—	—
52	11	1	1/11 = 0.0909	0.2928	3.951244	−1.22816
53	5	0	0/5 = 0.0000	0.2928	—	—
54	3	0	0/3 = 0.0000	0.2928	—	—
55	0	0	0/0 = –	–	–	—

Table 2.
One example of processing data for hazard plotting.

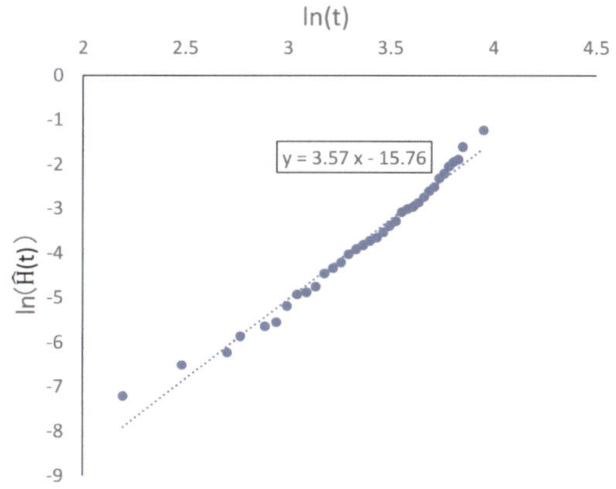

Figure 4.
One example of hazard plots.

performed since the failure would have occurred years later if no renewal was performed. In risk evaluation for the examination of maintenance and renewal plans, the use of failure rate characteristics based on the operational data without consideration of the preventive renewal effect is clearly an underestimate. Reference [8]

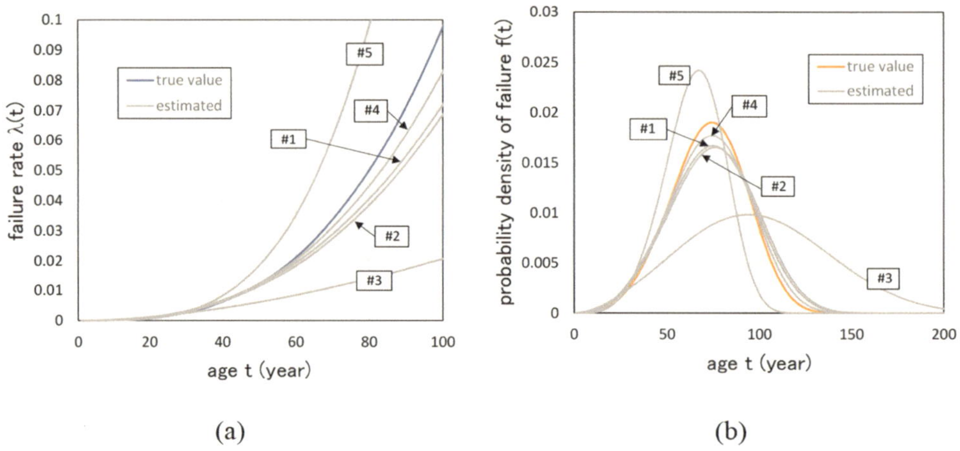

Figure 5.
Results of hazard analysis for 5 simulated data. (a) failure rate (b) probability density of failure.

analyzed operational performance data for transformers and shunt reactors and estimated failure rates assuming that equipment that was replaced before failure was the one that would fail 5 or 10 years later. In order to conduct such a study, it is necessary to investigate and accumulate the actual field data of not only operations and failures, but renewals with reasons.

An example of a survey of equipment operational data, including renewal data, is the questionnaire survey [9] conducted by "Investigating R&D Committee for Asset Management for Electric Power Equipment Based on Insulation Diagnosis" of IEE Japan. This is a survey of failure and renewal data in 10 years conducted on approximately 200 plant manufacturers and other companies. Reference [6] has tried to utilize the results of this survey to estimate failure rate characteristics by combining failure and renewal data. Although how to combine them should be examined according to the reasons for each renewal and the characteristics of the target equipment, the reasons for renewal were not investigated in the survey, therefore, the analysis was conducted by assuming that the failure should have occurred at +5 years after the year of each renewal [6]. The results for CV cables from 6 kV class to 60 kV class are shown in **Figures 6** and 7. For example, the failure rate at 30 years old is about 26 times higher than without taking the renewal results into account.

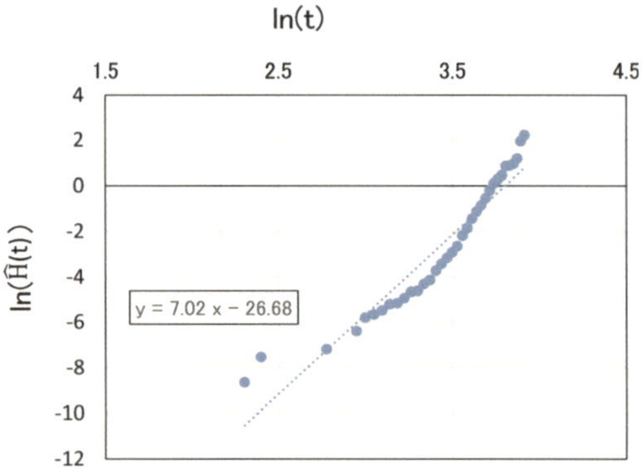

Figure 6.
Hazard plots of CV cables.

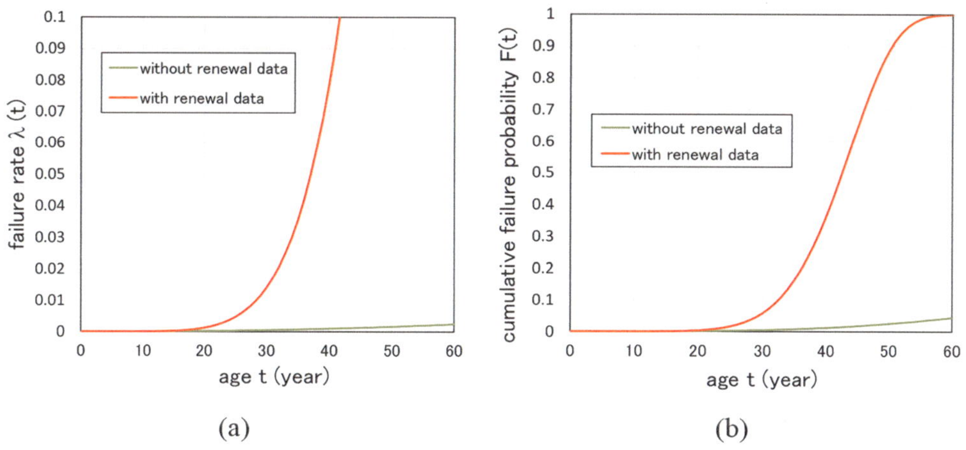

Figure 7.
Results of failure characteristics estimation of CV cable. (a) failure rate (b) Cumulative failure probability.

Considering that approximately 90% of the group is 6 kV class cables, and that diagnostic methods such as DC leakage current measurement would have been applied to this class cables, it can be assumed that most of the reasons for renewal are due to some kind of trouble, and it is highly appropriate to add up them when the failure rate is estimated.

4. Conclusion

In recent years, the asset management or managing assets for power transmission and distribution equipment has been actively examined. In order to optimize maintenance strategies, both risk and cost associated with operating the equipment should be considered, and both evaluated and compared in monetary values. CRIEPI is investigating managing assets support tools evaluating the cumulative cost in operation including statistically expected risk.

It is important to obtain failure rate characteristics of equipment for risk assessment. As one method, this chapter has presented a method for statistical analysis of actual equipment operation data in the field, as well as the necessary data and considerations for this method. For power transmission and distribution equipment with high reliability and low failure rates, it is necessary to accumulate actual data over a long period of time in order to accurately estimate failure rates. Among them, the information on equipment renewal, which has not necessarily been sufficiently investigated and analyzed in the past, is particularly important in the situation that the preventive maintenance is generally adopted, and must be investigated and accumulated together with the reasons for renewal.

Author details

Tsuguhiro Takahashi
Central Research Institute of Electric Power Industry (CRIEPI), Yokosuka-shi, Japan

*Address all correspondence to: shodai@criepi.denken.or.jp

References

[1] Ageing of the system – impact on planning, CIGRE TB176. 2000

[2] Takahashi T, Okamoto T. Development of asset management support tool for electric power apparatus. Int'l. Sympos. High Voltage Eng. (ISH), Paper No. T6-719. 2007

[3] Takahashi T, Okamoto T. Development of asset management support tools for oil immersed transformer. IEEE Transactions on Dielectrics and Electrical Insulation. 2016;**23**(3):1643-1648

[4] Takahashi T. Development of support program for managing assets by considering regular maintenance cost and statistically expected failure cost. In: Németh B, editor. Proceedings of the 21st International Symposium on High Voltage Engineering. ISH 2019. Lecture Notes in Electrical Engineering. Vol. 598. Springer, Cham; 2020. DOI: 10.1007/978-3-030-31676-1_22

[5] Takahashi T. Study of hierarchical support technique for managing assets - development of support program for substation equipment. 22nd International Symposium on High Voltage Engineering (ISH 2021). 2021. pp. 2101-2106. DOI: 10.1049/icp.2022.0463

[6] Takahashi T. Study of statistical failure analysis for long life equipment based on operation data. CRIEPI Report No. GD21025. 2022. [in Japanese]

[7] Guidelines for the use of statistics and statistical tools on life data. CIGRE TB706. 2017

[8] Picher P et al. Use of health index and reliability data for transformer condition assessment and fleet ranking. CIGRE session 45, A2-101. 2014

[9] Asset management for electric power equipment based on insulation diagnosis. Technical report of the Inst. of Electrical Engineers Japan, No. 1243. 2012. [in Japanese]

Section 2

Practical Examples

Chapter 4

A Study of Proportional Hazards Models: Its Applications in Prognostics

Chaoqun Duan and Lei Song

Abstract

Prognostics and health management technology is proposed to satisfy the requirements of equipment autonomous maintenance and diagnosis, which is a new technique relying on condition-based maintenance. It mainly includes condition monitoring, fault diagnostics, life prediction, maintenance decision-making, and spare parts management. As one of the most commonly used reliability statistical modeling methods, proportional hazards model (PHM) is widely used in the field of prognostics, because it can effectively combine equipment service age and condition monitoring information to obtain more accurate condition prediction results. In the past decades, PHM-based methods have been widely employed, especially since the twenty-first century. However, after the rapid development of PHM, there is no systematic review and summary particularly focused on it. Therefore, this chapter comprehensively summarizes the research progress of PHM in prognostics.

Keywords: proportional hazards models, prognostics, reliability engineering

1. Introduction

With the rapid development of science and technology, the integration, complexity, and intelligence of industrial systems have increased sharply. The traditional fault diagnostics and maintenance support technology is gradually difficult to adapt the new operation and maintenance requirements. As early as the 1970s, prognostics and health management technology first appeared. It achieves condition monitoring, fault prediction, and health management of complex industrial systems by processing and analyzing various operating data generated in the industrial process. This technology can predict the failure of the system before it happens and make effective maintenance decisions or suggestions in combination with the current working conditions. The main implementation steps are shown in **Figure 1**. In the figure, equipment prognostics and health management are divided into the processes of equipment condition monitoring, data acquisition, data processing, state prediction, and health management. Equipment prognostics is to predict the current or future state of equipment by using data acquisition and data processing technology based on condition monitoring information, including equipment failure rate, service

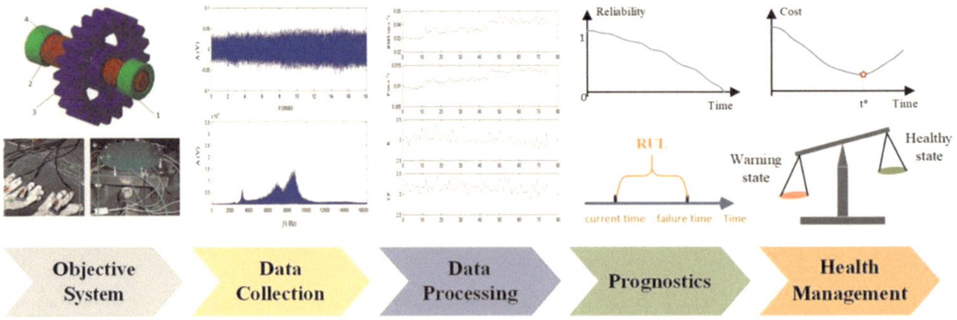

Figure 1.
Flowchart of equipment prognostics and health management.

reliability, remaining useful life (RUL), and other reliability indicators. The following health management measures are scheduled based on the predicted results of the component or system degradation state, such as reducing the frequency of monitoring, minimizing the number of maintenance downtimes, optimizing spare parts ordering and inventory management. In recent years, prognostic and health management has attracted extensive attention from industry and academia. There have also been several types of health prediction methods proposed, including physical model-based methods, data-driven methods, and hybrid model-based methods. In practical application, the physical model-based methods need to obtain the physical mechanism of the mechanical equipment degradation process. However, the physical processes for complex modern equipment are usually difficult to obtain. The data-driven method does not need to consider the kinematic principle of the equipment, instead relying on the data generated in the industrial dynamic process and extracting and processing the data to achieve the purpose of prediction. PHM is one of the commonly used models in data-driven methods. The hybrid model-based methods can combine the advantages of physical model and data-driven model to enhance the prediction accuracy, but designing the fusion mechanism between disparate models is a complex issue.

As early as 1972, David Cox [1] first proposed the PHM to characterize the effect of multiple factors on the mortality or failure rate at a given time. Initially, PHM was utilized in the biomedical domain to analyze the survival of cancer patients. The evolution of equipment failure rate has a certain similarity to human mortality, and PHM can better fit various risk processes over time. Therefore, PHM is also widely used in reliability modeling of industrial equipment. Compared with other reliability statistical models, PHM has the characteristics of universality, flexibility, and simplicity and can effectively incorporate information on equipment service age and condition monitoring data. This means that PHM can estimate the probability of equipment failure at any time in a given state and then evaluate the health state of the equipment. In addition, PHM is also suitable for dealing with censored data [2].

The remainder of this chapter is organized as follows. Firstly, Section 2 introduces the basic form of PHM, which is divided into four parts: baseline hazard function, link function, covariate process, and parameter estimation. Secondly, the research progress of PHM in prognostics is reviewed in Section 3, which mainly outlines the reliability evaluation and RUL prediction of PHM in various engineering fields. Finally, Section 4 summarizes the conclusions and discusses the advantages and current challenges of PHM.

2. Basic form of PHM

The PHM is used to describe the failure rate of equipment, which is related to time and covariates. It is usually represented by the product of two independent functions $h_0(t)$ and $\psi(\gamma Z_t)$. The failure rate function can be expressed as:

$$h(t, Z_t) = h_0(t)\psi(\gamma Z_t) \tag{1}$$

where $h_0(t)$ is the baseline hazard function, which is only related to the sequipment service time, and represents the failure rate of the equipment when it is not affected by covariates. $\psi(\gamma Z_t)$ is the link function, which is related to the value of the covariate Z_t and the covariate coefficient γ, indicating the influence of the covariate on the failure rate. Given the covariate Z_t, the failure rate function for service time t is defined as:

$$h(t, Z_t) = \lim_{\Delta t \to 0} \Pr(t < T < t + \Delta t | T > t, Z_t)/\Delta t \tag{2}$$

where T is the failure time of the system. According to the definition of equipment reliability, the equipment conditional reliability function can be derived as:

$$R(t|Z_u, 0 \leq u \leq t) = \Pr(T > t|Z_u, 0 \leq u \leq t) = \exp\left(-\int_0^t h(t, Z_u)du\right) \tag{3}$$

According to the definition of the remaining useful life X_t, $X_t = \{x_t : T - t|T > t, Z_t\}$. The probability density function of the RUL of the equipment at service time t can be expressed as:

$$f_{x_t}(x_t|Z_t) = \frac{f(t + x_t|Z_t)}{R(t|Z_t)} = h(t + x_t|Z_t)\frac{R(t + x_t|Z_t)}{R(t|Z_t)} \tag{4}$$

Next, the baseline hazard function, link function, covariate process, and parameter estimation of PHM will be introduced in detail.

2.1 Baseline hazard function

The baseline hazard function $h_0(t)$ can be modeled in various forms, including constant [3, 4], linear form [5], quadratic polynomial [6, 7], lognormal distribution [8, 9], and Weibull distribution [10–12]. Alternatively, following Cox's strategy, a distribution-free approach is employed to directly estimate the baseline hazard rate from historical failure event data [13, 14].

PHM can be classified into two types based on the form of the baseline hazard function: semi-parametric PHM and full-parametric PHM. When the baseline hazard function is not specified, the model in **Eq. (1)** is often referred to as a semi-parametric PHM. In fact, one of the major advantages of semi-parametric PHM is that there is no need to define a specific form for the baseline hazard function, which makes semi-parametric PHM more flexible. The original semi-parametric PHM can be fully parameterized by defining a specific form of the baseline hazard function. The Weibull distribution is often used to describe the baseline hazard function of PHM, because it covers various types of failure rates (increasing failure rate, constant failure rate, and decreasing failure rate) and can better fit the equipment degradation data. In the form of Weibull distribution, **Eq. (1)** can be further deduced as follows:

$$h(t, Z_t) = \frac{\beta}{\alpha}\left(\frac{t}{\alpha}\right)^{\beta-1}\psi(\gamma Z_t) \tag{5}$$

where $\alpha > 0$ is the scale parameter, and $\beta > 0$ is the shape parameter. The Weibull distSribution function is used as an example to demonstrate the fully parameterized method of semi-parametric PHM. The baseline hazard function of PHM can be simply extended to any other function form in addition to the Weibull distribution function.

2.2 Link function

The form of the link function $\psi(\gamma Z_t)$ depends on the given failure event data and must satisfy the condition $\psi(\gamma Z_t) > 0$. Cox proposed three link function forms, namely linear form $1 + \gamma Z$, inverse linear form $1/(1 + \gamma Z)$, and exponential function form $\exp(\gamma Z)$. However, for all possible values of Z, it is difficult to choose the coefficient γ in the linear and inverse linear forms to satisfy the above conditions. This criterion is better satisfied by the exponential function form. Moreover, an exponential function can also approximates the experimental data well [10]. Expressing $\psi(\gamma Z_t)$ in the form of an exponential function, **Eq. (1)** can be rewritten as the following form:

$$h(t, Z(t)) = h_0(t)\exp\left[\sum_{i=1}^{n}\gamma_i Z_i(t)\right] \tag{6}$$

where $h(t, Z(t))$ represents the failure rate at time t under the influence of the covariate vector $Z(t)$. The symbol γ_i,$i = 1,2, \dots ,n$, is the coefficient corresponding to the covariate $Z_i(t)$, indicating the degree of influence of each covariate on the failure rate, and n represents the number of covariates. Therefore, determining the covariate process and parameters of PHM is crucial for assessing equipment failure rates.

2.3 Covariate process

In reliability modeling, the environmental factors or self-degradation characteristics that affect the system failure rate are commonly referred to as covariates. Covariates are important factors in PHM, and the choice of covariates has a direct impact on the accuracy of reliability and life prediction. According to the internal and external factors affecting equipment failure, it can be divided into internal covariates and external covariates. Internal covariates include the system's own structural design [15], materials [16], degradation state characteristics [17], and so on. The current state of the system can be reflected according to the degradation characteristics. Whereas external covariates can usually be regarded as "risk factors" that can affect the failure time of the system, such as temperature [18, 19], humidity [20], weather conditions [21, 22], and other external operating environment.

During PHM modeling, a preliminary analysis of covariates should be performed to identify state indicators that have a significant impact on equipment failure rates. The covariates of PHM were determined by Vlok et al. [23] and Ghodrati et al. [24] based on the experience of maintenance technicians, which is extremely subjective. This could lead to the omission of other significant factors, as well as a high correlation between them. There are many methods to test the influence of covariates on the system failure rate. Most of the existing studies use methods such as P-value [25, 26], Wald test [27], likelihood ratio test [28], and score test [21]. When determining covariates, it is usually required that the correlation coefficient between the covariates be as small as possible. Therefore, Lin et al. [29], Carr et al. [30], and Chen et al. [31] employed principal component analysis (PCA) to analyze condition monitoring data and built PHM using principal components rather than the original covariates. This method is helpful to eliminate the collinearity between the original covariates and reduce the number of covariates in PHM. Makis et al. [32] used dynamic principal component analysis to reduce the dimensionality of the transmission oil data. Dynamic principal component analysis is an extension of the original PCA, which can achieve dimensionality reduction when the data have autocorrelation. Mazidi et al. [33] used several statistical techniques to decrease the dimension of the original monitoring data and select parameters, including PCA, Pearson, Spearman and Kendall correlation, mutual information, regression ReliefF, and decision trees. Ahmad et al. [34] used Failure Mode Effect and Criticality Analysis (FMECA) to identify external covariates that may affect the failure rate of transmission belts in cutting process system. Another key reason they use FMECA is that it can classify censored and uncensored data. Then, a statistical analysis of censored and uncensored time-to-failure data was performed by applying Failure Time Modeling (FTM) based on PHM considering the effects of external covariates. In order to investigate the impact of different covariates, Kabir et al. [35] and Kabir et al. [36] stratified the data according to the material type of the water mains and whether they had previously failed to establish distinct PHMs. They identified significant covariates in different models using the Bayesian model averaging (BMA) method. Based on the assumption that the heavier the operational use of components, the higher the probability of component failure, Verhagen et al. [37] used Extreme Value Analysis (EVA) and Maximum Difference Analysis (MDA) techniques to identify the operational factors that lead to the high failure rate of aircraft components. Wu et al. [28] used the Z test to investigate the effect of time-varying environmental covariates on the failure rate of wind turbine components, and the likelihood ratio test was used to find the best combination of covariates. Li et al. [38] and Thijssens et al. [39] used the Akaike Information Criterion (AIC) to choose covariates. In addition, the software SPSS [40, 41] and EXAKT [42–44] can also be applied to identify critical covariates affecting the equipment failure rate.

2.4 Parameter estimation

Parameter estimation is an important step in PHM modeling, including baseline hazard function parameter estimation and link function parameter estimation. The covariate most closely connected to system failure should have a higher weight in the link function, and the corresponding covariate coefficient γ_i should be greater. The covariates having a weak link to failure should be given less weight, and the corresponding covariate coefficient γ_i should be smaller. The model will be fairly close to reality if only covariates related to system failure are included. Therefore, the accuracy of the model parameter estimation has a considerable influence on the calculation result of the total failure rate of the objective equipment. Generally, the parameters of PHM are estimated by partial likelihood function [45, 46] or maximum likelihood function [8, 47], or related software programs, such as SPSS [40, 41], EXAKT [48, 49], SYSTAT [3], survival package for R [50], coxphfit function of

Matlab [25], and so on. The likelihood estimation function formula of PHM parameter estimation is simple, and the maximization process is robust. However, the method based on classical likelihood estimation may suffer from slow convergence. Moreover, the maximum likelihood method cannot quantify the uncertainty in model predictions and field data. Uncertainties can be found in the whole process of calculation and modeling, including those resulting from test data and model parameters. For the lack of sufficient experimental or field data, Zuashkiani et al. [51] used expert knowledge to compensate and developed a method combining expert knowledge and statistical data to estimate the parameters of PHM. Considering the uncertainties in model predictions and field data, Jiang et al. [52] built a Bayesian network to represent nonlinear PHM based on historical inspection data. When updating the distribution function to estimate the model parameters in Bayesian networks, it is necessary to calculate the marginal function, which usually requires the high-dimensional integration problem on the prior distribution. Therefore, they used the Markov Chain Monte Carlo technique to solve the difficulties in parameter estimation.

The framework for equipment prognostics and health management based on PHM is shown in **Figure 2**, which includes data acquisition, PHM modeling, prognostics, and health management. Firstly, state indications connected to equipment failure, such as temperature, current, or vibration signals of the objective equipment, are collected using manual operation, sensors, or specific test tools. Then, P-value, principal component analysis or expert experience is used to determine the covariates that have a significant impact on the equipment failure rate. At the same time, maximum likelihood estimation, Bayesian update or SPSS, EXAKT, and other software are used to estimate the parameters for the PHM modeling. Finally, the reliability indicators such as equipment reliability or RUL are estimated according to the established PHM to achieve the purpose of state prediction. The basic structural form of PHM (containing the baseline hazard function and link function), as well as the covariate determination method and commonly used parameter estimation methods, is all introduced in this section. Next, as shown in **Figure 2**, we will concentrate on the prediction and evaluation of PHM in the domains of cutting tools, bearings, water

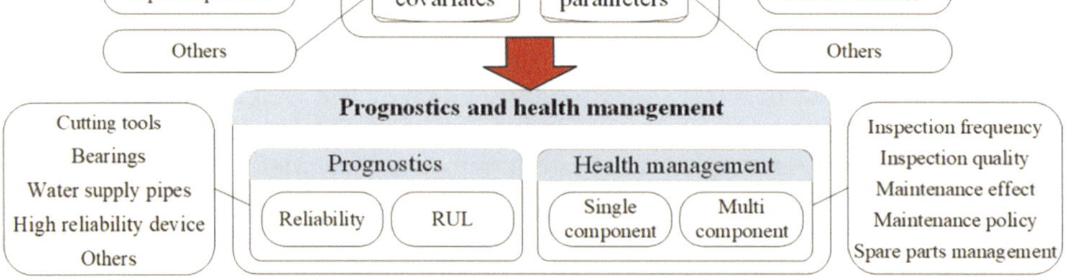

Figure 2.
Framework diagram of equipment prognostics and health management based on PHM.

supply pipes, and high-reliability devices and primarily review the covariate indicators selected in various literatures, as well as reliability estimation and RUL prediction based on PHM.

3. Research progress of PHM in prognostics

In equipment failure prediction, reliability and RUL are two key health indicators. Reliability refers to the ability of a product to accomplish a specific function under specified conditions and within a specified time, indicating the probability that the product will fail to occur within a certain period of time. RUL refers to the continuous operating time of the equipment from the present moment to the occurrence of a potential failure. A potential risk to an effective forecasting system lies in accurately assessing reliability, RUL, and other relevant reliability indicators. Therefore, in order to avoid failures, accurate prediction of reliability and RUL through quantitative methods based on the current state of the machine and operating history is crucial for making preventive maintenance (PM) decisions. In PHM, the RUL of the equipment can be derived from the relevant reliability function, as shown in **Eq. (4)**. Initially, Bendell [53] pointed out that PHM offers a lot of potential in the field of reliability assessment. According to the existing research, PHM has been widely employed in failure data analysis, reliability assessment, and life prediction in various fields, including hardware and software [54, 55]. For example, valves [56, 57], aircraft cargo doors [58], mining loader cables [59–61], distribution network cables [41, 62], printed circuit boards [63], mobile handsets [64], electrical appliances [65], automotive air-conditioning compressors [66], and so on. In addition, Barker et al. [67] applied PHM to describe the instantaneous rate of recovery of an electric power system after an outage and the likelihood of recovery occurs prior to a given point in time. Mohammad et al. [68] used PHM and Markov chains to analyze the reliability of load-sharing systems with a k-out-of-n structure. Zhao et al. [69] described a task reliability modeling method based on the Quality State Task Network (QSTN), used the WPHM to estimate the reliability of the cylinder head manufacturing system, and then described the overall operation state of the system. In various application examples, PHM is widely used in reliability assessment and life prediction in the domains of cutting tools, bearings, water supply pipes, and high-reliability devices. **Table 1** gives an introductory summary of PHM in various application domains, regarding issues and failures, common measures, common covariates, and example data.

3.1 Cutting tools

In industrial manufacturing, the cost of consumables such as cutting tools cannot be ignored. The estimation error of cutting tool reliability or residual life may result in a large amount of production loss. On the one hand, overestimating tool reliability or residual life can lead to substandard parts being produced, as well as poor surface quality and machine damage. On the other hand, the underestimate of tool reliability or residual life may reduce the overall productivity and raise production cost, since insufficient tool use and shutdown loss result from needless frequent tool replacement. Therefore, effectively predicting tool reliability and RUL can help production managers to develop better tool replacement strategies, improve production planning, and increase production efficiency. At present, some scholars have used PHM to combine machining time and different working conditions to estimate tool reliability

Application domains	Issues and failures	Common measures	Common covariates	Example data
Cutting tools	Tool wear, tool fracture, poor surface finish, blade cracking	Vibration, tool wear value, cutting parameters	Cutting speed, feed rate and depth of cut [70], root mean square and peak [8, 9], cutting speed and the feed rate [49, 71], tool wear, cutting speed and feed rate [72], cutting speed [50], the logarithm of cutting speed [73]	A CNC lathe FTC-20, FAIR FRIEND Group Taiwan [8], a six-axis Boehringer NG 200 [49], Gamma process simulation [50], a CNC SOMAB "UNIMAB 450" lathe [73]
Bearings	Outer-race, inner-race, roller, and cage failures	Vibration, oil debris, acoustic emission	Natural logarithm of root mean square and kurtosis [74, 75], kurtosis factor and crest factor [76], kurtosis [77], standard deviation, root mean square, and root amplitude sequences [78]	Experimental data of bearings 6205-2RS (SKF) from Case Western Reserve University [76], the prognostic data repository contributed by Intelligent Maintenance System, University of Cincinnati [77, 79]
Water supply pipes	Corrosion, leaking of joints, main barrel and line valves, blockage, break	Physical parameters of pipes, pressure, surrounding environment	Diameter, length, corrosivity, soil stability, internal pressure and the percentage of the pipe covered with low development land [80], material, diameter, length, vintage, soil type and the number of previous failures [81], length, diameter, pipe material, soil resistivity, soil resistivity, freezing index, rain deficit [82]	A pipe database collected in Laramie, Wyoming [83], the water distribution system serving the western part of the province of Ferrara [84], a dataset on pipe breakage from the city of Limassol (Cyprus) [85], the failure database of water distribution network in the City of Calgary, Alberta, Canada [35, 36, 82]

Table 1.
Introductory summarization of PHM in various application domains.

or RUL. Mazzuchi et al. [70] used PHM to evaluate the reliability of machine tools and used a full Bayesian method to compare the prior and posterior distributions of the parameters involved in the model to reflect tool aging and the importance of each covariate. In an automated manufacturing system, it is usually necessary for the same tool to cut multiple parts from different materials with different cutting parameters in order to save the space of tool magazine and avoid frequent tool changes while increasing production efficiency. In this situation, Liu et al. [86] derived a formula for calculating tool reliability under various cutting conditions with random machining time. Ding et al. [8] and Ding et al. [9] used PHM to analyze tool wear reliability by extracting the root mean square and peak of time domain indicators from the tool vibration signal as covariates. Cutting speed was chosen as a covariate of PHM by Equeter et al. [50], and the Mean Up Time of cutting tools was calculated using the integral of the reliability function. However, they only investigated the effect of cutting speed as a covariate on tool life. Shaban et al. [49] presented the reliability and RUL curves of tool cutting titanium metal matrix composites (Ti-MMCs) under various cutting speeds and feed rates. In the work of Equeter et al. [73], the authors converted the cutting speed of the tool to logarithmic form and used it as a covariate of PHM to predict the average available time of the tool. The logarithmic conversion of cutting speed data can provide more accurate prediction results, as demonstrated by a numerical case. Aramesh et al. [71] proposed a cutting tool life prediction model that took into account the influence of cutting parameters, machining time, and different tool wear stages (initial wear zone, steady wear zone, and rapid wear zone). The model provides very good estimates of tool life and critical points at which changes of states take place, as well as can calculate each between-states transition time. Aramesh et al. [72] developed a model for estimating the RUL of the worn tool under various cutting situations purely based on the actual wear of the tool, regardless of its usage history. This is a significant advantage of this model over other models in practical applications.

3.2 Bearings

Bearings are widely employed in a variety of areas as the essential components of rotating machinery. Due to the severe operating environment, bearing failure is one of the most common causes of machine failure, so it is necessary to perform reliability assessment and RUL prediction on bearings to prevent unexpected failures or accidents. Ding et al. [87] extracted the kurtosis and the root mean square in the bearing vibration signal as covariates reflecting the bearing operating state to evaluate the reliability of the bearing. To evaluate the reliability of bearing on site, Ding et al. [76] extracted the kurtosis factor and the crest factor as the covariates of PHM. The evaluation results can reflect the trend of failure occurrence and development. In some cases, it is very difficult to collect data from the actual system. Therefore, Leturiondo et al. [25] employed a physical model to generate synthetic data related to bearing degradation in order to fit the PHM and then estimate the bearing reliability further. The PHM presented by Liao et al. [74] takes into account both hard failure and multiple degradation features. The model is able to predict the mean RUL of a component based on online degradation information. Liao et al. [75] further compared the approach of Liao et al. [74] with the logistic regression model, demonstrating that the estimated RUL value based on PHM is closer to the actual life through a bearing test.

In recent research work, some scholars have combined PHM with artificial intelligence algorithms to estimate bearing reliability and predict RUL. Caesarendra et al. [77] used reliability theory and PHM to estimate the failure degradation of bearings and regarded it as a target vector. At the same time, combined with the kurtosis in the bearing vibration signal, they trained the support vector machine and established the life prediction model. The trained support vector machine is then utilized to predict the failure time of an individual bearing. Combining a neural network and PHM, Wang et al. [78] proposed a three-phase prognostic algorithm for bearings reliability evaluation and life prediction, which included feature selection, feature prediction, and RUL prediction. To begin, the most useful time-dependent features of vibration signals were extracted. Then, the feed-forward neural network is established as an identification model to predict the future features trends. Finally, PHM is used to estimate the reliability and RUL of the bearing. Qiu et al. [79] proposed an ensemble RUL prediction model by combining feature extraction, genetic algorithm, support vector regression, and WPHM. In this approach, genetic algorithm and signal feature extraction techniques are used to construct an effective health indicator. Secondly, support vector regression is used to predict the future development of the system operation behavior. Finally, RUL prediction is implemented using the WPHM prediction function.

3.3 Water supply pipes

The failure of water supply pipes usually affects other nearby infrastructure, which can lead to catastrophic consequences, so a large number of articles have been published to study the break risk process of water supply pipes. Kleiner et al. [88] outlined the application of various statistical models to water mains degradation, of which PHM is one of the most commonly used statistical models for estimating break failure of water mains. PHM was initially used by Jeffrey [7] to model the failure rate of water distribution system. Andreou et al. [89] and Andreou et al. [90] introduced the concept of early and late stages of water distribution system failures and used PHM to predict the deterioration of water distribution system in early stages with fewer breaks. They distinguished the different stages of pipe breaks based on a fixed number of failures, which only applies to the specific scenario considered, and did not explicitly describe the method used to identify the different stages. Park [91] and Park [92] developed a methodology to assess and track changes in the hazard functions between water main breaks by using PHM. As the number of pipe breaks increases, the critical points when the hazard function changes into different functional forms can be obtained to distinguish different stages of pipe failure. Park et al. [93] and Park et al. [94] divided cast iron 6-inch pipes into seven groups according to the break history of the water distribution system and constructed different PHMs for each group to estimate the reliability of the pipes. When there are only brief maintenance records, Le Gat et al. [95] discussed the efficiency of WPHM in fault prediction of water networks. Alvisi et al. [84] further investigated the model proposed by Le Gat et al. [95], pointing out that WPHM can exploit the information available on both the characteristics of the pipes in which breakages occur and their age to make the prediction results more stable and reliable. Instead of calculating the expected number of failures for a group of pipes, Clark et al. [96] and Karaa et al. [80] used PHM to calculate the probability of a pipe breaking or leaking for each pipe. Vanrenterghem-Raven et al. [97] created a simple prioritization index based on the ratio of pipe failure rates to determine which pipes should be replaced first. PHM was used by Fuchs-Hanusch et al. [81] to estimate the years when the failures occur with a defined probability. Moreover, they proposed a whole of life cost calculation method due to the long lifetime of water supply pipes. Christodoulou [85] used a 5-year dataset to study the impact of several risk factors on pipe failure, such as pipe material, diameter, and accident type. Regardless of the quality and quantity of data utilized in the model, there is inherent uncertainty when predicting the failure of water pipes. In order to explain the variability of these unknown factors, Clark et al. [83] incorporated a shared frailty into the PHM to account for the unspecified variability affecting the pipe breaks. Kabir et al. [35] and Kabir et al. [36] developed a Bayesian framework for predicting water main failure in the face of uncertainties. The proposed Bayesian Weibull proportional hazards model (BWPHM) is applied in this study to develop survival curves and predict water main failure rates. The results of their case indicated that the predicted 95% uncertainty bounds of the proposed BWPHMs capture effectively the observed water main failures. Applying the receiver operating characteristics curve, Debón et al. [98] compared PHM and generalized linear model for evaluating the risk of failure in water supply networks. Kimutai et al. [82] compared the predicted effects of Cox PHM, WPHM, and Poisson model in the break of cast iron, ductile iron, and plastic water pipes. The results recommended that a combined model should be used according to the rate of degradation and material type of the system. Xie et al. [99] used PHM to study the blockage risk of vitrified clay wastewater pipes and identified the pipes with the highest risk of failure due to blockage. In a cost-constrained environment, targeted inspection, plan maintenance, and replacement programs can be carried out to reduce the serious consequences caused by blockage.

3.4 High-reliability device

The PHM is a very popular tool in reliability theory and applications, which can be used to simulate the impact of another environment on the reliability of a baseline environment. For long-life and high-reliability devices (such as some electronic components [100, 101], etc.), it is difficult to obtain their failure data in a short time. Accelerated testing is a method for decreasing the life of high reliability devices or accelerating their performance degradation. The results of the tests are obtained in a shorter period of time under an accelerated stress environment, and PHM is then utilized to predict the failure behavior of the device under various operating conditions [102, 103]. There are generally two approaches to comprehensively utilize the failure data at various stress levels. The first is to convert data collected under high stress into data collected under normal operating conditions in order to expand the sample size and improve the accuracy of parameter estimation, reliability assessment, and life prediction [104]. Another approach is to establish the relationship between stress environment and lifetime by using acceleration models such as Arrhenius models, power law models, and exponential models [5]. Comparing PHM with the accelerated failure time model, Newby [105] pointed out the greatest advantage of PHM is that it does not need to specify a baseline failure rate, and it can quantitatively analyze the impact of each covariate on the total failure rate. Elsayed et al. [6] and Finkelstein [106] generalized the covariates of a single stress type to two stress types, allowing for the collection of a large number of failure time data in a short period of time. PHM was introduced into dealing with accelerated degradation test data by Chen et al. [107], who proposed a model based on the proportional degradation hazards model. They plotted the reliability curves of carbon-film resistors at normal stress condition based on the proposed model. To explore the reliability trend of Metal-Oxide-Semiconductor Field-Effect Transistors, Zheng et al. [108] conducted an accelerated degradation test with temperature as the accelerated stress. They established a PHM with the degradation trend and temperature as covariates, in which the degradation trend is defined by the Wiener process, because the degradation trend contains more information than the degradation state. The reliability results predicted in this reference are closer to the real scenario than the PHM with only temperature or only the deterioration state as covariate.

3.5 Others

In addition to the above domains, PHM can also be used in many other industries such as batteries, pumps, wind turbines, and so on. Some scholars also use PHM to describe the probability of hard failure of equipment. Hard failure generally refers to the sudden failure of equipment due to hidden manufacturing defects, excessive loads, or other stresses. When a hard failure occurs, the degradation signal tends to exhibit different values, such as the resistance value of a lead-acid battery in an automobile. Zhou et al. [109] proposed a two-stage approach, with an offline modeling stage based on historical data and an online prediction stage based on the degradation signal of each individual unit. This method is suitable for predicting the RUL of batteries without a powerful computing platform such as vehicle microcontrollers; however, the prediction results are relatively conservative. The variance of resistance between different batteries becomes larger as time increases, and there are noticeable individual differences among units. Therefore, in the RUL prediction framework of Man et al. [110], the authors applied the Wiener process with drift to characterize the degradation path of the battery resistance. However, because of the Markov nature of the Wiener process, their prediction methods rely on the most recent observational information. In comparison to Zhou et al. [109], the prediction accuracy is relatively low when current observations deviate from the degenerate path.

In the hard failure prediction method, the above studies do not consider potential change points in condition monitoring signals. However, change point detection and equipment degradation modeling are interrelated, which directly affects the accuracy of residual life prediction. You et al. [111] detected the change point in advance through the statistical process control method and divided the life cycle of the equipment into two zones: the stable zone and the degradation zone. This study was limited to the assumption that the change point was fixed and did not consider the impact of other factors on the change point during equipment operation. Son et al. [112] extended the method of Zhou et al. [109] to predict the RUL for individual units with considering the change point in condition monitoring signals, where the change point is captured based on the concordance correlation coefficient (CCC). Although this method improves the accuracy of RUL prediction, it also increases the complexity and computational burden of the model.

Because of the complexity of modern mechanical systems and the diversity of failure modes, it is necessary to consider the competition and interaction between different failure modes when analyzing the failure of the whole system. Zhang et al. [113] proposed a mixture Weibull proportional hazards model (MWPHM) to predict the failure of a high-pressure water descaling pump with two failure modes of sealing ring wear and thrust bearing damage. The system failure probability density is obtained by proportionally accumulating the probability density of multiple failure modes. Compared with traditional WPHM, MWPHM can provide more detailed life information, and its failure probability distribution is closer to the actual distribution. In this model, the prediction of failure time depends on the choice of reliability threshold. However, they assumed the reliability threshold was fixed, ignoring the fact that the reliability of the repaired system may change at the moment of failure.

In recent years, machine learning methods have been continuously developed. These theories and methods have been widely employed in engineering applications and scientific fields to address complex problems. In reliability engineering, some scholars compared the reliability and RUL prediction results based on PHM with the prediction results of neural network [103] and random forests method [114]. Li et al. [114] investigated the effect of failure time data with heavy-tailed behavior on the RUL prediction error. The results showed that the RUL prediction method based on PHM can make more accurate mechanical failure predictions than random forests. Izquierdo et al. [115] proposed a reliability model based on a dynamic artificial neural network by combining the neural network model and dynamic PHM concept. The model combines the benefits of neural networks for analyzing unknown interactions between environmental variables with the benefits of PHM for integrating dynamic operational environments. Mazidi et al. [116] created three neural network models to simulate the normal behavior of three features of wind turbine rotor speed, gearbox temperature, and generator winding temperature. Deviation signals are defined and calculated as accumulated time series of differences between neural network predictions and actual measurements. These signals are then used to develop a health condition model for each considered feature of wind turbine in order to perform anomaly detection. By combining autoregressive moving average model, PHM, and support vector machine, Tran et al. [117] proposed a three-stage method for estimating low methane compressor performance degradation and RUL. The method only uses the normal operation condition of the machine to create an identification model to recognize the dynamic system behavior and does not need data of whole machine life. Chen et al. [118] designed a deep learning structure called merged-long-short term memory (M-LSTM) network for health index modeling, which they subsequently integrated with PHM to predict the RUL of an automobile. However, since the repaired automobile cannot be recovered to its original state, the authors only consider the first maintenance record of automobiles, which makes it difficult to construct the health index of the automobile.

PHM has a certain application in prognostics and has achieved considerable results, but there is still a lack of a lot of research work to extend the prognostics scheme based on PHM.

4. Conclusions

As one of the most commonly used statistical models, PHM has received extensive attention and has been applied in a variety of domains. This chapter has summarized the research progress of prognostics based on PHM, focusing on the baseline hazard function, link function, covariate process, and parameter estimation methods. Data analysis, reliability estimation, and RUL prediction based on PHM are systematically discussed.

According to the review and research of PHM in this chapter, the advantages of PHM in the field of prognostics are summarized. Compared with other statistical methods, the main advantages of PHM are as follows.

1. PHM does not require assumptions about the nature or form of the baseline hazard function, and any type of distribution function can be used as the baseline hazard function. Therefore, PHM can be applied to reliability analysis of equipment in a variety of engineering domains, and it has the characteristics of universality, flexibility, and simplicity.

2. PHM directly models the failure rate, which has strong interpretability. The relationship between failure rate and condition monitoring information has been established, allowing the condition monitoring information to be used more effectively to update the equipment state. Furthermore, the influence of various covariates on the total failure rate may be easily assessed.

3. PHM can better simulate the influence of multiple internal and external degradation information on equipment failure, including environmental factors, aging factors, and degradation factors. Therefore, PHM is applicable when the failure of equipment is related to multiple influencing factors.

4. Compared with other data-driven methods, PHM can achieve good modeling results with a limited amount of degraded data. The accuracy of equipment condition prediction will improve as the amount of gathered event data and condition monitoring data grows.

Although PHM has made significant development in the field of prognostics and health management and has many advantages mentioned above, there are still a few aspects that need to be studied further. The current challenges facing PHM are listed below in order to point out the development direction for researchers.

1. Proportionality assumption. The application of PHM needs to satisfy the proportionality assumption, which has a fixed model form. It can only be used if the influence of the degradation process on the failure rate satisfies the link function, which is a fairly severe requirement.

2. Determine covariates. PHM can take into account the effects of multiple covariates on the failure rate of equipment at the same time. However, the results of parameter estimation will be biased if a relevant covariate is omitted or the accuracy of covariate measurement varies. Furthermore, because multiple covariates are associated with the same equipment degradation process, there may be correlations between them, which might influence the accuracy of the prediction results if not treated properly.

3. Data fusion. It is a challenge for data-specific fusion methods, such as the fusion between vibration signals, current signals, and oil signals. Complex systems usually involve multidimensional covariate processes. To assure the accuracy of system state prediction, more research into how to properly integrate diverse forms of data and adopt a more reasonable combination structure deserves further study.

4. Calculation difficulty. For a covariate process affected by stochastic degradation, a stochastic process must be used to describe the degradation process of covariates, which increases the calculation burden. The calculation of PHM becomes extremely difficult when the degradation process of the system incorporates multiple covariates. Calculating high-dimensional data presents a number of challenges.

5. Data problems. It is difficult to obtain event data and condition monitoring data simultaneously in practical applications. Especially for high-reliability systems and crucial equipment, they are not allowed to run to failure. This leads to a small number of failed samples, posing a major barrier to data-driven methods.

Combining the above advantages and disadvantages, PHM is suitable for imperfect observation systems with small degraded data samples and can make better use of condition monitoring information in various dimensions. The data fusion and model calculation problems of PHM can be greatly solved when combined with other methods (such as Bayesian iteration, artificial intelligence, and so on), offering it irreplaceable theoretical value and application prospect in the field of prognostics and health management in complex systems.

Author details

Chaoqun Duan* and Lei Song
School of Mechatronic Engineering and Automation, Shanghai University, Shanghai, China

*Address all correspondence to: chaoqun.duan@hotmail.com

References

[1] Cox DR. Regression models and life-tables. Journal of the Royal Statistical Society: Series B (Methodological). 1972; **34**(2):187-202

[2] Wightman DW, Bendell A. The practical application of proportional hazards modelling. Reliability Engineering. 1986;**15**(1):29-53

[3] Ghodrati B, Kumar U. Operating environment-based spare parts forecasting and logistics: A case study. International Journal of Logistics. 2005; **8**(2):95-105

[4] Samrout M, Châtelet E, Kouta R, et al. Optimization of maintenance policy using the proportional hazard model. Reliability Engineering and System Safety. 2009;**94**(1):44-52

[5] Pang Z, Hu C, Si X, et al. Life prediction approach by integrating nonlinear accelerated degradation model and hazard rate model. 2018 Prognostics and System Health Management Conference (PHM-Chongqing). IEEE. 2018:392–398

[6] Elsayed EA, Hao Z. Design of PH-based accelerated life testing plans under multiple-stress-type. Reliability Engineering and System Safety. 2007; **92**(3):286-292

[7] Jeffrey LA. Predicting Urban Water Distribution Maintenance Strategies: A Case Study of New Haven, Connecticut. Massachusetts Institute of Technology; 1985

[8] Ding F, He Z. Cutting tool wear monitoring for reliability analysis using proportional hazards model. International Journal of Advanced Manufacturing Technology. 2011; **57**(5–8):565-574

[9] Ding F, Zhang L, He Z. On-line monitoring for cutting tool wear reliability analysis. In: Proceedings of the World Congress on Intelligent Control and Automation. IEEE. 2011. pp. 364-369

[10] Jóźwiak IJ. An introduction to the studies of reliability of systems using the Weibull proportional hazards model. Microelectronics Reliability. 1997;**37**(6): 915-918

[11] Jardine AKS, Anderson PM, Mann DS. Application of the Weibull proportional hazards model to aircraft and marine engine failure data. Quality and Reliability Engineering International. 1987;**3**(2):77-82

[12] Jardine AKS, Ralston P, Reid N, et al. Proportional hazards analysis of diesel engine failure data. Quality and Reliability Engineering International. 1989;**5**(3):207-216

[13] Baxter MJ, Bendell A, Manning PT, et al. Proportional hazards modelling of transmission equipment failures. Reliability Engineering and System Safety. 1988;**21**(2):129-144

[14] Love CE, Guo R. Using proportional hazard modelling in plant maintenance. Quality and Reliability Engineering International. 1991;**7**(1):7-17

[15] Krivtsov VV, Tananko DE, Davis TP. Regression approach to tire reliability analysis. Reliability Engineering and System Safety. 2002;**78**(3):267-273

[16] Bendell A, Walley M, Wightman D, et al. Proportional hazards modelling in reliability analysis—An application to brake discs on high speed trains. Quality and Reliability Engineering International. 1986;**2**(1):45-52

[17] Tang D, Yu J, Chen X, et al. An optimal condition-based maintenance policy for a degrading system subject to the competing risks of soft and hard failure. Computers and Industrial Engineering. 2015;**83**:100-110

[18] Li L, Ma D, Li Z. Cox-proportional hazards modeling in reliability analysis-a study of electromagnetic relays data. IEEE Transactions on Components, Packaging and Manufacturing Technology. 2015;**5**(11):1582-1589

[19] Jóźwiak IJ. Use of concomitant variables for reliability exploration of microcomputer systems. Microelectronics Reliability. 1992;**32**(3): 341-344

[20] Izquierdo J, Crespo A, Uribetxebarria J, et al. Assessing the impact of operational context variables on rolling stock reliability. A real case study. Safety and Reliability–Safe Societies in a Changing World CRC Presstime. 2018. pp. 571-578

[21] Ezzeddine W, Schutz J, Rezg N. Cox regression model applied to Pitot tube survival data. 2015 International Conference on Industrial Engineering and Systems Management (IESM). IEEE. 2016. pp. 168-172

[22] Ezzeddine W, Schutz J, Rezg N. Test for additive interaction in proportional hazard model applied to Pitot sensors reliability and survivability. IFAC-Papers OnLine. Vol. 49(2). 2016. pp. 1-5

[23] Vlok PJ, Coetzee JL, Banjevic D, et al. Optimal component replacement decisions using vibration monitoring and the proportional-hazards model. Journal of the Operational Research Society. 2002;**53**(2):193-202

[24] Ghodrati B, Kumar U. Reliability and operating environment-based spare parts estimation approach: A case study in Kiruna Mine, Sweden. Journal of Quality in Maintenance Engineering. 2005;**11**(2):169-184

[25] Leturiondo U, Salgado O, Galar D. Estimation of the Reliability of Rolling Element Bearings Using a Synthetic Failure Rate. Springer International Publishing; 2016. pp. 99-112

[26] Tracht K, Goch G, Schuh P, et al. Failure probability prediction based on condition monitoring data of wind energy systems for spare parts supply. CIRP Annals-Manufacturing Technology. 2013;**62**(1):127-130

[27] Lin D, Wiseman M, Banjevic D, et al. An approach to signal processing and condition-based maintenance for gearboxes subject to tooth failure. Mechanical Systems and Signal Processing. 2004;**18**(5):993-1007

[28] Wu F, Zhou Y, Liu J. Modelling the effect of time-dependent covariates on the failure rate of wind turbines. In: Asset Intelligence through Integration and Interoperability and Contemporary Vibration Engineering Technologies. Springer; 2019. pp. 727-734

[29] Lin D, Banjevic D, Jardine AKS. Using principal components in a proportional hazards model with applications in condition-based maintenance. Journal of the Operational Research Society. 2006;**57**(8):910-919

[30] Carr MJ, Wang W. A case comparison of a proportional hazards model and a stochastic filter for condition-based maintenance applications using oil-based condition monitoring information. Journal of Risk and Reliability. 2008;**222**(1):47-55

[31] Chen Z, Ren J, Zhang Y, et al. Maintenance decision of door system

based on PHM-assisted RCM. 2017 36th Chinese Control Conference (CCC). IEEE. 2017. pp. 7433-7437

[32] Makis V, Wu J, Gao Y, et al. An application of DPCA to oil data for CBM modeling. European Journal of Operational Research. 2006;**174**(1): 112-123

[33] Mazidi P, Bertling Tjernberg L, Sanz Bobi MA. Wind turbine prognostics and maintenance management based on a hybrid approach of neural networks and a proportional hazards model. Proceedings of the Institution of Mechanical Engineers, Part O: Journal of Risk and Reliability. 2017;**231**(2):121-129

[34] Ahmad R, Kamaruddin S, Azid IA, et al. Failure analysis of machinery component by considering external factors and multiple failure modes – A case study in the processing industry. Engineering Failure Analysis. 2012;**25**: 182-192

[35] Kabir G, Tesfamariam S, Sadiq R. Predicting water main failures using Bayesian model averaging and survival modelling approach. Reliability Engineering and System Safety. 2015; **142**:498-514

[36] Kabir G, Tesfamariam S, Loeppky J, et al. Predicting water main failures: A Bayesian model updating approach. Knowledge-Based Systems. 2016;**110**: 144-156

[37] Verhagen WJC, de Boer LWM. Predictive maintenance for aircraft components using proportional hazard models. Journal of Industrial Information Integration. 2018:23-30

[38] Li Z, Zhou S, Choubey S, et al. Failure event prediction using the Cox proportional hazard model driven by frequent failure signatures. IIE Transactions. 2007;**39**(3):303-315

[39] Thijssens OWM, Verhagen WJC. Application of extended cox regression model to time-on-wing data of aircraft repairables. Reliability Engineering & System Safety. 2020;**204**:107136

[40] Kobbacy K, Fawzi B, Percy D, et al. A full history proportional hazards model for preventive maintenance scheduling. Quality and Reliability Engineering International. 1997;**13**(4): 187-198

[41] Tang Z, Zhou C, Wei J, et al. Analysis of significant factors on cable failure using the cox proportional hazard model. IEEE Transactions on Power Delivery. 2014;**29**(2):951-957

[42] Tian Z, Liao H. Condition based maintenance optimization for multi-component systems using proportional hazards model. Reliability Engineering & System Safety. 2011;**96**(5):581-589

[43] Jardine AKS, Banjevic D, Wiseman M, et al. Optimizing a mine haul truck wheel motors' condition monitoring program Use of proportional hazards modeling. Journal of Quality in Maintenance Engineering. 2001;7(4): 286-302

[44] Wong EL, Jefferis T, Montgomery N. Proportional hazards modeling of engine failures in military vehicles. Journal of Quality in Maintenance Engineering. 2010;**16**(2): 144-155

[45] Cox DR. Partial likelihood. Biometrika. 1975;**2**:269-276

[46] Tail M, Yacout S, Balazinski M. Replacement time of a cutting tool subject to variable speed. Proceedings of the Institution of Mechanical Engineers,

Part B: Journal of Engineering Manufacture. 2010;**224**(3):373-383

[47] Love CE, Guo R. Application of weibull proportional hazards modelling to bad-as-old failure data. Quality & Reliability Engineering International. 1991;7(3):149-157

[48] Jardine AKS, Joseph T, Banjevic D. Optimizing condition-based maintenance decisions for equipment subject to vibration monitoring. Journal of Quality in Maintenance Engineering. 1999;**5**(3):192-202

[49] Shaban Y, Yacout S. Predicting the remaining useful life of a cutting tool during turning titanium metal matrix composites. Proceedings of the Institution of Mechanical Engineers Part B Journal of Engineering Manufacture. 2018;**232**(4):681-689

[50] Equeter L, Letot C, Serra R, et al. Estimate of Cutting Tool Lifespan through Cox Proportional Hazards Model. Ifac Papersonline. Vol. 49(28). 2016. pp. 238-243

[51] Zuashkiani A, Banjevic D, Jardine AKS. Estimating parameters of proportional hazards model based on expert knowledge and statistical data. Journal of the Operational Research Society. 2009;**60**(12):1621-1636

[52] Jiang X, Yuan Y, Liu X. Bayesian inference method for stochastic damage accumulation modeling. Reliability Engineering and System Safety. 2013; **111**:126-138

[53] Bendell A. Proportional hazards modelling in reliability assessment. Reliability Engineering. 1985;**11**(3): 175-183

[54] Bendell A, Wightman DW, Walker EV. Applying proportional hazards modelling in reliability. Reliability Engineering & System Safety. 1991;**34**(1):35-53

[55] Ansell JI, Philipps MJ. Practical aspects of modelling of repairable systems data using proportional hazards models. Reliability Engineering & System Safety. 1997;**58**(2):165-171

[56] Lindqvist B, Molnes E, Rausand M. Analysis of SCSSV performance data. Reliability Engineering & System Safety. 1988;**20**(1):3-17

[57] Booker J, Campbell K, Goldman AG, et al. Applications of Cox's proportional hazards model to light water reactor component failure data. Los Alamos Scientific Lab; 1981

[58] Leitao ALF, Newton DW. Proportional hazards modelling of aircraft cargo door complaints. Quality and Reliability Engineering International. 1989;**5**(3):229-238

[59] Kumar D, Westberg U. Proportional hazards modeling of time-dependent covariates using linear regression: A case study [mine power cable reliability]. IEEE Transactions on Reliability. 1996; **45**(3):386-392

[60] Kumar D. Proportional hazards modelling of repairable systems. Quality and Reliability Engineering International. 1995;**11**(5):361-369

[61] Kumar D, Klefsjö B, Kumar U. Reliability analysis of power transmission cables of electric mine loaders using the proportional hazards model. Reliability Engineering & System Safety. 1992;**37**(3):217-222

[62] Nemati HM, Sant'anna A, Nowaczyk S, et al. Reliability evaluation of power cables considering the restoration characteristic. International

Journal of Electrical Power and Energy Systems. 2019;**105**:622-631

[63] Drury MR, Walker EV, Wightman DW, et al. Proportional hazards modelling in the analysis of computer systems reliability. Reliability Engineering & System Safety. 1988; **21**(3):197-214

[64] Tiwari A, Roy D. Estimation of reliability of mobile handsets using Cox-proportional hazard model. Microelectronics Reliability. 2013;**53**(3): 481-487

[65] Mendes AC, Fard N. Reliability modeling for appliances using the Proportional Hazard Model. 2013 Proceedings Annual Reliability and Maintainability Symposium (RAMS). IEEE, 2013. pp. 1-6

[66] Landers TL, Kolarik WJ. Proportional hazards analysis of field warranty data. Reliability Engineering. 1987;**18**(2):131-139

[67] Barker K, Baroud H. Proportional hazards models of infrastructure system recovery. Reliability Engineering & System Safety. 2014;**124**: 201-206

[68] Mohammad R, Kalam A, Amari SV. Reliability of load-sharing systems subject to proportional hazards model. 2013 Proceedings Annual Reliability and Maintainability Symposium (RAMS). IEEE. 2013. pp. 1-5

[69] Zhao Y, He Y, Chen Z, et al. Big operational data oriented health diagnosis based on weibull proportional hazards model for multi-state manufacturing system. 2018 Prognostics and System Health Management Conference (PHM-Chongqing). IEEE. 2018. pp. 444-449

[70] Mazzuchi TA, Soyer R. Assessment of machine tool reliability using a proportional hazards model. Naval Research Logistics. 1989;**36**(6):765-777

[71] Aramesh M, Shaban Y, Yacout S, et al. Survival life analysis applied to tool life estimation with variable cutting conditions when machining titanium metal matrix composites (Ti-MMCs). Machining Science and Technology. 2016;**20**(1):132-147

[72] Aramesh M, Attia M, Kishawy H, et al. Estimating the remaining useful tool life of worn tools under different cutting parameters: A survival life analysis during turning of titanium metal matrix composites (Ti-MMCs). CIRP Journal of Manufacturing Science and Technology. 2016;**12**:35-43

[73] Equeter L, Ducobu F, Rivière-Lorphèvre E, et al. An analytic approach to the cox proportional hazards model for estimating the lifespan of cutting tools. Journal of Manufacturing and Materials Processing. 2020;**4**(1):27

[74] Liao H, Qiu H, Lee J, et al. A predictive tool for remaining useful life estimation of rotating machinery components. International Design Engineering Technical Conferences and Computers and Information in Engineering Conference. Vol. 47381. 2005

[75] Liao H, Zhao W, Guo H. Predicting remaining useful life of an individual unit using proportional hazards model and logistic regression model. RAMS'06 Annual Reliability and Maintainability Symposium. IEEE. 2006. pp. 127-132

[76] Ding F, He Z. Formalization of reliability model for assessment and prognosis using proactive monitoring mechanism. 2010 Prognostics and

System Health Management Conference. IEEE. 2010. pp. 1-6

[77] Caesarendra W, Widodo A, Yang BS. Combination of probability approach and support vector machine towards machine health prognostics. Probabilistic Engineering Mechanics. 2011;**26**(2): 165-173

[78] Wang L, Zhang L, Wang XZ. Reliability estimation and remaining useful lifetime prediction for bearing based on proportional hazard model. Journal of Central South University. 2015;**22**(12):4625-4633

[79] Qiu G, Gu Y, Chen J. Selective health indicator for bearings ensemble remaining useful life prediction with genetic algorithm and Weibull proportional hazards model. Measurement. 2020;**150**:107097

[80] Karaa FA, Marks DH. Performance of water distribution networks: Integrated approach. Journal of Performance of Constructed Facilities. 1990;**4**(1):51-67

[81] Fuchs-Hanusch D, Kornberger B, Friedl F, et al. Whole of life cost calculations for water supply pipes. Water Asset Management International. 2012;**8**(2):19-24

[82] Kimutai E, Betrie G, Brander R, et al. Comparison of statistical models for predicting pipe failures: Illustrative example with the City of Calgary water main failure. Journal of Pipeline Systems Engineering and Practice. 2015;**6**(4): 04015005

[83] Clark RM, Carson J, Thurnau RC, et al. Condition assessment modeling for distribution systems using shared frailty analysis. Journal-American Water Works Association. 2010;**102**(7):81-91

[84] Alvisi S, Franchini M. Comparative analysis of two probabilistic pipe breakage models applied to a real water distribution system. Civil Engineering and Environmental Systems. 2010;**27**(1): 1-22

[85] Christodoulou SE. Water network assessment and reliability analysis by use of survival analysis. Water Resources Management. 2011;**25**(4):1229-1238

[86] Liu H, Makis V. Cutting-tool reliability assessment in variable machining conditions. IEEE Transactions on Reliability. 1997;**45**(4): 573-581

[87] Ding F, He Z, ZI Y, ET AL. Reliability assessment based on equipment condition vibration feature using proportional hazards model Journal of Mechanical Engineering. 2009;45(12):89–94

[88] Kleiner Y, Rajani B. Comprehensive review of structural deterioration of water mains: Statistical models. Urban water. 2001;**3**(3):131-150

[89] Andreou SA, Marks DH, Clark RM. A new methodology for modelling break failure patterns in deteriorating water distribution systems: Theory. Advances in Water Resources. 1987;**10**(1):2-10

[90] Andreou SA, Marks DH, Clark RM. A new methodology for modelling break failure patterns in deteriorating water distribution systems: Applications. Advances in Water Resources. 1987; **10**(1):11-20

[91] Park S. Identifying the hazard characteristics of pipes in water distribution systems by using the proportional hazards model: 1 Theory. KSCE Journal of Civil Engineering. 2004;**8**(6):663-668

[92] Park S. Identifying the hazard characteristics of pipes in water distribution systems by using the proportional hazards model: 2 Applications. KSCE Journal of Civil Engineering. 2004;**8**(6):669-677

[93] Park S, Kim JW, Newland A, et al. Survival analysis of water distribution pipe failure data using the proportional hazards model. World Environmental and Water Resources Congress 2008: Ahupua'A. 2008. pp. 1-10

[94] Park S, Jun H, Agbenowosi N, et al. The proportional hazards modeling of water main failure data incorporating the time-dependent effects of covariates. Water Resources Management. 2011; **25**(1):1-19

[95] Le Gat Y, Eisenbeis P. Using maintenance records to forecast failures in water networks. Urban water. 2000; **2**(3):173-181

[96] Clark RM, Goodrich JA. Developing a data base on infraestructure needs. Journal-American Water Works Association. 1989;**81**(7):81-87

[97] Vanrenterghem-Raven A, Eisenbeis P, Juran I, et al. Statistical modeling of the structural degradation of an urban water distribution system: Case study of New York City. World Water & Environmental Resources Congress 2003. 2003. pp. 1-10

[98] Debón A, Carrión A, Cabrera E, et al. Comparing risk of failure models in water supply networks using ROC curves. Reliability Engineering & System Safety. 2010;**95**(1):43-48

[99] Xie Q, Bharat C, Nazim Khan R, et al. Cox proportional hazards modelling of blockage risk in vitrified clay wastewater pipes. Urban Water Journal. 2016:1-7

[100] Elsayed EA, Chan CK. Estimation of thin-oxide reliability using proportional hazards models. IEEE Transactions on Reliability. 1990;**39**(3): 329-335

[101] Zhao S, Makis V, Chen S, et al. Health assessment method for electronic components subject to condition monitoring and hard failure. IEEE Transactions on Instrumentation and Measurement. 2018;**68**(1):138-150

[102] Dale CJ. Application of the proportional hazards model in the reliability field. Reliability Engineering. 1985;**10**(1):1-14

[103] Luxhoj JT, Shyur HJ. Comparison of proportional hazards models and neural networks for reliability estimation. Journal of Intelligent Manufacturing. 1997;**8**(3):227-234

[104] Prasad PVN, Rao KRM. Reliability models of repairable systems considering the effect of operating conditions. Annual Reliability and Maintainability Symposium 2002 Proceedings (Cat No 02CH37318). IEEE. 2002. pp. 503-510

[105] Newby M. Perspective on Weibull proportional-hazards models. IEEE Transactions on Reliability. 1994;**43**(2): 217-223

[106] Finkelstein M. On dependent items in series in different environments. Reliability Engineering & System Safety. 2013;**109**:119-122

[107] Chen HT, Yuan HJ. Reliability assessment based on proportional degradation hazards model. 2010 IEEE 17th International Conference on Industrial Engineering and Engineering Management. IEEE. 2010. pp. 958-962

[108] Zheng H, Kong X, Xu H, et al. Reliability analysis of products based on

proportional hazard model with degradation trend and environmental factor. Reliability Engineering & System Safety. 2021;**216**:107964

[109] Zhou Q, Son J, Zhou S, et al. Remaining useful life prediction of individual units subject to hard failure. IIE Transactions. 2014;**46**(10): 1017-1030

[110] Man J, Zhou Q. Prediction of hard failures with stochastic degradation signals using wiener process and proportional hazards model. Computers & Industrial Engineering. 2018;**125**: 480-489

[111] You M-Y, Li L, Meng G, et al. Two-zone proportional hazard model for equipment remaining useful life prediction. Journal of Manufacturing Science and Engineering. 2010;**132**(4): 041008

[112] Son J, Zhang Y, Sankavaram C, et al. RUL prediction for individual units based on condition monitoring signals with a change point. IEEE Transactions on Reliability. 2014;**64**(1):182-196

[113] Zhang Q, Hua C, Xu G. A mixture Weibull proportional hazard model for mechanical system failure prediction utilising lifetime and monitoring data. Mechanical Systems and Signal Processing. 2014;43(1-2):103–112

[114] Li Z, Kott G. Predicting Remaining Useful Life Based on the Failure Time Data with Heavy-Tailed Behavior and User Usage Patterns Using Proportional Hazards Model//2010 Ninth International Conference on Machine Learning and Applications. IEEE. 2010: 623-628

[115] Izquierdo J, Marquez AC, Uribetxebarria J. Dynamic Artificial Neural Network-based reliability considering operational context of assets. Reliability Engineering & System Safety. 2019;**188**(AUG.):483-493

[116] Mazidi P, Du M, Bertling Tjernberg L, et al. A health condition model for wind turbine monitoring through neural networks and proportional hazard models. Proceedings of the Institution of Mechanical Engineers, Part O: Journal of Risk and Reliability. 2017;**231**(5):481-494

[117] Tran VT, Hong TP, Yang BS, et al. Machine performance degradation assessment and remaining useful life prediction using proportional hazard model and support vector machine. Mechanical Systems & Signal Processing. 2012;**32**:320-330

[118] Chen C, Liu Y, Sun X, et al. An integrated deep learning-based approach for automobile maintenance prediction with GIS data. Reliability Engineering & System Safety. 2021;**216**:107919

Chapter 5

Pull-Type Security Patch Management in Intrusion Tolerant Systems: Modeling and Analysis

Junjun Zheng, Hiroyuki Okamura and Tadashi Dohi

Abstract

In this chapter, we introduce a stochastic framework to evaluate the system availability of an intrusion tolerant system (ITS), where the system undergoes patch management with a periodic vulnerability checking strategy, i.e., pull-type patch management. In particular, a composite stochastic reward net (SRN) is developed to capture the overall system behaviors, including vulnerability discovery, intrusion tolerance, and reactive maintenance operations. Furthermore, two kinds of availability criteria, the interval availability and the steady-state availability of the system, are formulated by applying the phase-type (PH) approximation to solve the Markov regenerative process (MRGP) model derived from the composite SRN. Numerical experiments are conducted to investigate the effects of the vulnerability checking interval on the system availability.

Keywords: intrusion tolerance system, security patch management, vulnerability checking, interval availability, steady-state availability, stochastic reward net, Markov regenerative process, phase expansion

1. Introduction

Computer systems face an increased number of security threats, which exploit the system's potential vulnerability to breach computer security, eventually causing possible damages such as information leakage and economic losses. Software testing is important for ensuring a program's quality, but it is acceptable that perfect software is impossible to achieve. For example, software vulnerabilities are discovered and disclosed continuously, even though developers carefully execute software testing in the development phase [1]. Online vulnerability databases such as MITRE Corporation's Common Vulnerabilities and Exposures (CVE) list[1] and Open Source Vulnerability Database (OSVDB)[2] have reported a vast number of vulnerabilities for recent years. According to CVE, 69,417 vulnerabilities were discovered in web applications over the years 1999–2015 [2]. Due to the existence of vulnerabilities, the risk to cyber

---

1 http://www.cve.mitre.org

2 http://www.osvdb.org

security becomes more significant, and the tricks of attacks also become cleverer and more sophisticated [3]. That means how to guarantee computer security against malicious attacks is a challenging task.

Computer security generally has three attributes; that is, confidentiality, integrity, and availability (CIA) [4]. Two typical techniques, i.e., intrusion detection [5] and intrusion tolerance [6], have been developed and well studied to protect the CIA. Intrusion detection is traditionally used to prevent intrusion as a proactive barrier by monitoring the system behavior. For example, misuse detection is to find the detection signature and anomaly detection is to predict the system's anomaly by comparing normal profiles. Nevertheless, unfortunately, intrusion detection is not still efficient enough to prevent recent and sophisticated malicious attacks. On the other hand, intrusion tolerance is practical to keep the correct services even under attack by masking intrusion based on fault-tolerant techniques for software faults. Some well-known intrusion tolerant systems (ITSs) are, for instance, the SITAR (scalable intrusion tolerant architecture) [7], a concrete ITS architecture using COTS (commercial-off-the-shelf) distributed servers, the BFT-WS [8], a Byzantine fault-tolerant framework for web services providers, and the virtual machine (VM) based ITS, a multistage ITS in virtualized computing environments [9–11].

However, there is no doubt that the most efficient way to ensure computer security is to apply a patch to fix the vulnerable system before a malicious attack occurs. The problem now in patch management from the user's perspective is when to apply the patch because the system may stop while the patch is applied. Even for ITSs, it is essential to decide on an appropriate patch management strategy. Some literature studies have considered such a security patch management from the user's perspective. For example, Kansal et al. [12] presented a generalized framework to identify the optimal patch applying strategy and its minimum cost when the level of system reliability is retained. Uemura et al. [13] focused on typical DoS (denial-of-service) attacks for SITAR and formulated the optimal security patch management policy via semi-Markov models in terms of system availability. In [13], a push-type patch management was considered; that is, the vulnerability information was pushed to a client whenever a new vulnerability was discovered. In the push-type patch management, a patch can be applied just after release. But in fact, for open software projects, such as Apache httpd server, the users need to check the vulnerability information by themselves; that is, pull-type patch management. Therefore, this chapter considers the security patch management of SITAR architecture and discusses the pull-type patch management strategies.

In this chapter, based on two availability measures, we reveal the effect of the number of checking on the system availabilities. More specifically, we develop a composite stochastic reward net (SRN) model [14] with the following four submodules: a vulnerability model to describe the vulnerability discovery process, an intrusion tolerance model to capture the system behaviors under reactive defense strategies after the occurrence of a security failure, a clock model to control the periodic checking interval, and a maintenance model to adopt the preventive and corrective actions for security threats. Also, the phase-type (PH) expansion approach is applied to analyze the Markov regenerative process (MRGP) derived from the SRN to evaluate two kinds of system availabilities. The stationary analysis of MRGP is generally achieved by employing an embedded Markov chain (EMC) approach based on Markov renewal theory [15–18]. Despite this, it is relatively difficult for transient cases. Besides, for the situation where the state in MRGP has multiple competitive transitions timed with generally distributed firing time (GEN transition), it is difficult to analyze the MRGP through Markov renewal equations since it is difficult to use the discretization and supplementary variable method [19]. Therefore, in this chapter, we seek to bridge this gap by developing the solution with PH expansion [19, 20], which is to replace general distributions in MRGP with approximate PH distributions and reduce the original MRGP to an approximate continuous-time Markov chain (CTMC). The accuracy of PH approximation has been validated in [20]. In particular, this chapter utilizes PH expansion of MRGP based on the Kronecker representation.

The remaining part of this chapter is organized as follows. In Section 2, we introduce an overview of an ITS and describe its composite SRN. Section 3 presents the performance analysis through MRGP analysis and PH-expansion CTMC analysis. In particular, the system's interval availability and steady-state availability under patch management are formulated. In Section 4, we present evaluation results. The conclusion and future work are given in Section 5.

2. Intrusion-tolerant system

2.1 System architecture

Consider an intrusion-tolerant architecture as in **Figure 1**, which is the SITAR architecture [7]. In this figure, the part within the denoted box is regarded as an intrusion tolerant architecture that enables us to build intrusion-tolerant servers out of the existing intrusion vulnerable servers $S_1, S_2, \ldots, S_i$. The architecture consists of five critical components: proxy server, acceptance monitor, ballot monitor, adaptive

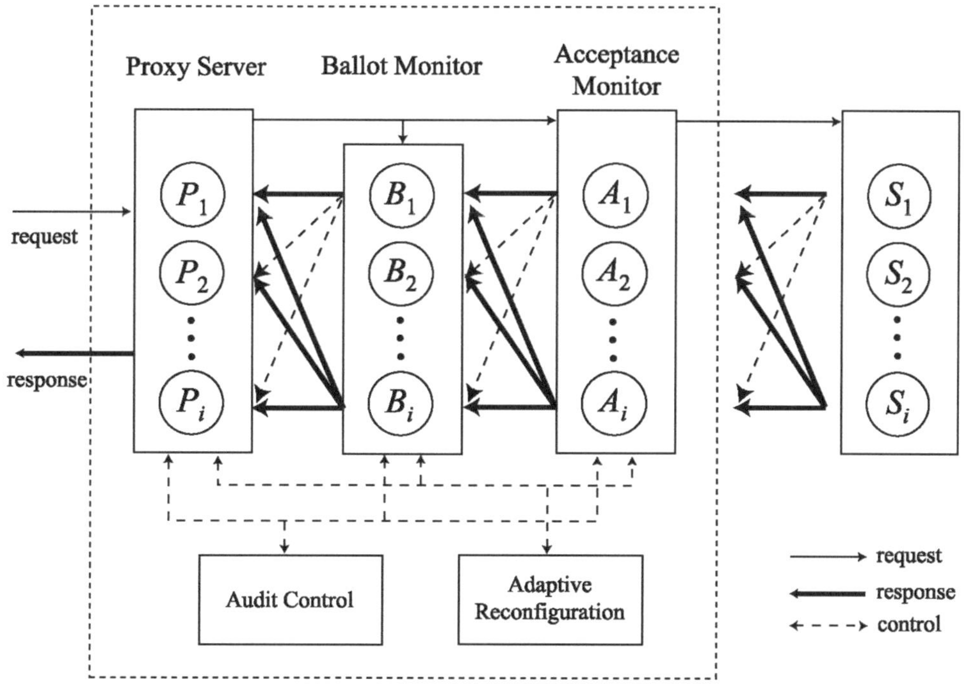

Figure 1.
Intrusion-tolerant architecture.

reconfiguration module, and audit control module. P_i, B_i, and A_i in the functional blocks are the logical functions to be executed to satisfy a given service request.

The proxy servers act as public access points for the services provided. When a request from remote client arrives at one of the proxy servers depending on the service needs, the proxy servers forward the request to one or more COTS servers based on the current intrusion-tolerant strategy. After receiving the COTS servers' responses, the acceptance monitors apply certain validity check to these responses and then forward them along with a check result indication to the ballot monitors. Besides, the acceptance monitors detect the signs of compromised servers and produce intrusion triggers for the adaptive reconfiguration module. The ballot monitors make a final response by either a simple majority voting or Byzantine agreement process and then forward the final response to the proxy servers to be delivered to the remote client.

The audit control module monitors the behaviors of all the other components in the system, by verifying their audit logs. When intrusion is detected, the corresponding information will be sent to the adaptive reconfiguration module. The adaptive reconfiguration module receives intrusion trigger information from all other modules, evaluates the threats, the tolerance objectives, and the cost/performance impact, and finally generates new configurations for the system.

2.2 System behavior

2.2.1 Intrusion tolerance scheme

The system becomes vulnerable once the vulnerability in servers S_1, S_2, … , S_i is disclosed. In this state, the system may encounter security threats that exploit discovered vulnerabilities. When a malicious attack arrives, the system moves to the active attack state and attempts to detect the intrusion threat. If the threat is detected successfully, the system begins to diagnose the detected threat and then tries to mask the compromised part; otherwise, security failure occurs and then a recovery process, namely corrective maintenance, is conducted. The system becomes normal again after the recovery ends.

For the case where the intrusion threat is detected successfully and the masking of compromised parts succeeds, the system can continually provide services to users after a minor fix in the background. Once the masking fails, several corrective inspections are tried in parallel with services. If a fatal system error is inspected, the system fails and becomes unavailable. In such a case, a recovery operation is executed to fix a fatal system error. The system goes back to the normal again after the completion of the recovery operation. If a fatal system error is not found, the system can keep servicing with a degraded performance if the attack's damage is not so large, or move to a fail-secure state otherwise, in which the system stops servicing to users. In either case, the system becomes normal after removing the system secure errors.

On the other hand, the system applies security patches if preventive maintenance (i.e., security patch application) is triggered before the attack. After completing the preventive maintenance, the state becomes normal.

2.2.2 Periodic vulnerability checking strategy

Maintenance strategies aim to prevent malicious attacks by executing the security patch application. This chapter considers pull-type patch management with a periodic

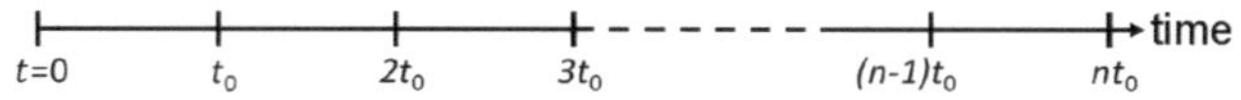

Figure 2.
Periodic vulnerability checking points.

vulnerability checking strategy. **Figure 2** illustrates the periodic checking points for discovered vulnerabilities. The length of one checking interval is given by t_0, and the time points t_0, $2t_0$, …, nt_0 are checking points for deciding whether to implement patches or not. At these checking points, if discovered vulnerabilities exist, the system stops providing services and executes a patch application. Otherwise, the system continues to provide services. The pull-type patch management with a periodic vulnerability checking strategy is described as follows.

Apply the security patch if discovered vulnerabilities exist in the system at the checking points. The length of the checking interval is denoted by t_0 (>0).

2.3 Stochastic reward net

The SRN is a highly representative model, consisting of: place P, represented by circle; transition T, represented by box; directed arcs, connecting places and transitions; and token(s). A transition is enabled if all of its input places have at least one token. When a transition is enabled, it may be fired to remove one token from each input place and create one token at each output place. Places may be marked by an integer number of tokens. The overall state of a system is represented by a vector consisting of the markings on each place. In SPN, there may exist the following types of transitions; (i) IMM transition (immediate, i.e., they fire in zero time); (ii) EXP transition (timed with exponentially distributed firing time); and (iii) GEN transition (timed with generally distributed firing time). In general, the IMM transition, EXP transition, and GEN transition are often expressed by a thin black bar, a white box, and a thick black bar, respectively. When more than two transitions are enabled simultaneously, guard functions are added to these transitions to control the firing sequence. A transition with a guard function occurs when the value of the guard function is evaluated to be true. The SRN can capture common characteristics of computer systems such as concurrency, synchronization, and sequencing, so it is widely used for stochastic modeling.

In this chapter, we present an SRN with the following submodules for the aforementioned ITS:

1. Vulnerability model, which depicts the vulnerability discovery process.

2. Intrusion tolerance model, which determines the system operation after a security threat occurs.

3. Clock model, which controls the checking interval.

4. Maintenance model, which describes the preventive and corrective actions for security threats.

Figure 3 depicts the composite SRN of the ITS with the pull-type patch management described in 2.2.2.

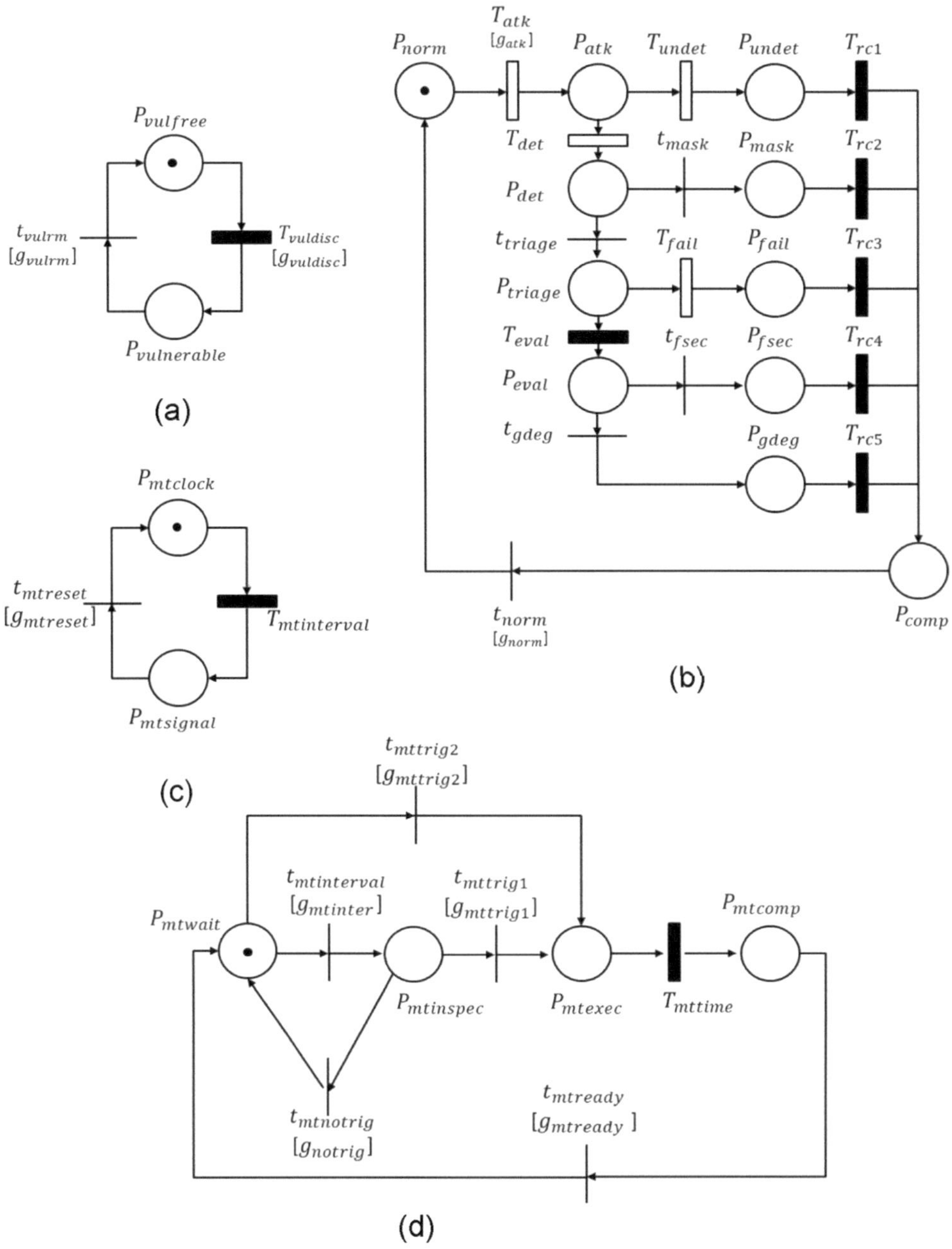

Figure 3.
Composite SRN for the ITS (a) Vulnerability model, (b) intrusion tolerance model, (c) clock model, and (d) maintenance model.

2.3.1 Vulnerability model

Figure 3a depicts an SRN of the vulnerability discovery process. As in **Figure 3a**, the model has two place ($P_{vulfree}$ and $P_{vulnerable}$), one IMM transition (t_{vulrm}) and one EXP transition ($T_{vuldisc}$). A token in $P_{vulfree}$ denotes that the system is vulnerability-free, i.e., no vulnerability has been discovered. When $T_{vuldisc}$ fires, one token is removed from $P_{vulfree}$

and put in $P_{vulnerable}$, which means that the vulnerability is discovered, and the system becomes vulnerable. Once the value of the guard function of t_{vulrm} is true (i.e., the system is under patch application), the system returns the vulnerability-free state immediately.

2.3.2 Intrusion tolerance model

Figure 3b presents an SRN of the intrusion tolerance model, which determines the system operation after a security threat occurs. In this figure, GEN transitions with the generally distributed firing times (represented by thick black bars) are used. Each place and corresponding transition represent the status of progress of an intrusion tolerant process and are given as **Table 1**.

Node	Description
P_{norm}	The system is in a normal state.
P_{atk}	Threat has occurred in the system. The system attempts to detect the threat.
P_{undet}	Threat cannot be detected. The security failure occurs due to the attack and the system is forced to undergo recovery processes.
P_{det}	Threat has been detected. The system begins to diagnosis the detected threat.
P_{mask}	The compromised part is being masked. Concretely, the system provides services to users, though minor errors causing threat are being fixed in the background.
P_{triage}	Threat triage state. Several corrective inspections are tried in parallel with services.
P_{fail}	The system fails and starts a recovery operation to fix a fatal system error.
P_{eval}	The damage of attack is being evaluated.
P_{fsec}	The system becomes fail-secure. The system stops servicing to users and applies recovery operation.
P_{gdeg}	The system keeps servicing while the quality of service is degraded.
P_{comp}	The recovery operation is completed.
T_{atk}	The system is attacked by adversary.
T_{undet}	The threat is undetected.
T_{det}	The threat is detected.
t_{mask}	The compromised part is masked.
t_{triage}	Threat triage begins.
T_{fail}	The system fails.
T_{eval}	The damage of attack is evaluated.
t_{fsec}	The system becomes fail-secure.
t_{gdeg}	The system degrades.
T_{rc1}	The system is in recovery process regarding detection failure.
T_{rc2}	The system is in recovery process regarding masking.
T_{rc3}	The system is in recovery process regarding system failure.
T_{rc4}	The system is in recovery process regarding fail-secure.
T_{rc5}	The system is in recovery process regarding graceful degradation.

Table 1.
Places and transitions in SPN for intrusion tolerance model (see ***Figure 3b****).*

2.3.3 Clock and maintenance models

In this chapter, the security patch application is regarded as the maintenance action. **Figure 3d** and **c** describe the maintenance model and its clock model. As in **Figure 3c**, the clock model controls the checking interval; that is, if a checking point is reached, the transition $T_{mtinterval}$, corresponding to the checking interval t_0, fires, then the token in $P_{mtclock}$ is removed, and a token is put into $P_{mtsignal}$. Upon confirmation that the maintenance model has received the signal of reaching a checking point (i.e., $\#(P_{mtinspec}) = 1$), the clock is reset with transition $t_{mtreset}$ immediately. On the other hand, from **Figure 3d**, we see that the maintenance model contains four places, one GEN transition, five IMM transitions, and one token in P_{mtwait}, indicating that the system is waiting for a maintenance operation. Besides, a token in $P_{mtinspec}$ represents that the system is checking whether to execute a patch application; once there exists discovered vulnerabilities at the checking point (i.e., the guard function $g_{mttring1}$ is true), the system performs patch application; otherwise, the system continues to wait for the next checking point. A token in P_{mtexec} means that the system is carrying out the maintenance, and the time spent is given by transition T_{mttime}. A token in P_{mtcomp} says that a maintenance is completed, and then the system goes back to the normal state with transition t_{norm} in **Figure 3b** and becomes ready for the next maintenance chance through transition $t_{mtready}$. Note that transition $t_{mttrig2}$ indicates the maintenance triggered due to a security threat.

In these above SRNs, the guard functions are shown in **Table 2**, which determine the enabled timing and are given by the interrelationships between the transition and the corresponding places. A marking of composite SRN is given by a vector that represents the number of tokens for all the places and provides the state of ITS. Actually, the composite SRN can be described by the underlying stochastic process, called MRGP [21], and analyzed by using MRGP analysis based on Markov renewal theory [15, 16]. The MRGP is one of the favored techniques for modeling system behavior with non-Markovian processes, can adequately represent more complex

	Guard function
$g_{vuldisc}$	$\#(P_{mtwait}) = 1$
g_{vulrm}	$\#(P_{mtexec}) = 1$
g_{atk}	$\#(P_{vulnerable}) = 1$
g_{norm}	$\#(P_{mtcomp}) = 1$
$g_{mtreset}$	$\#(P_{mtinspec}) = 1\ \#(P_{mtexec}) = 1$
$g_{mtinter}$	$\#(P_{mtsignal}) = 1$
g_{notrig}	$(\#(P_{norm}) = 0\ \#(P_{vulfree}) = 1)\ \&\&\ \#(P_{mtclock}) = 1$
$g_{mttrig1}$	$\#(P_{norm}) = 1\ \&\&\ \#(P_{vulnerable}) = 1\ \&\&\ \#(P_{mtclock}) = 1$
$g_{mttrig2}$	$\#(P_{comp}) = 1$
$g_{mtready}$	$\#(P_{norm}) = 1$

Table 2.
Enabling functions in the composite SRN.

software intrusion tolerant process and maintenance actions, and has been successfully applied in several modeling analyses [16–19].

3. Performance analysis

The performance criteria of interest in this chapter are the interval availability and the steady-state availability of the system, which require the state probabilities of MRGP derived according to the analysis of composite SRN described in 2.3 by using JSPetriNet software package[3]. The MRGP model of ITS is depicted in **Figure 4**. In this figure, the solid lines denote the GEN transitions, whereas the dashed ones denote EXP transitions. In particular, all states except S^G_{mtint} have two competitive GEN transitions. In such a case, it is difficult to obtain the state probabilities of MRGP through Markov renewal equations because it is hard to use the discretization and supplementary variable method [19]. This chapter, therefore, considers the solution with phase-type (PH) expansion for analyzing the MRGP model of the ITS. Also, in this chapter, we utilize the PH expansion of MRGP based on the Kronecker representation.

3.1 PH approximation

The phase expansion, alternatively PH approximation, is the technique by using PH distribution, which is defined as the probability distribution of the absorbing time in a CTMC with absorbing states. The PH distribution is practical, since it can approximate any probability distribution with high precision. To take benefit from this property, an approximate CTMC can be obtained by replacing probability distribution with PH distributions. Without loss of generality, the infinitesimal generator $\mathbf{Q}$ of CTMC is assumed to be partitioned as follows:

$$\mathbf{Q} = \begin{pmatrix} \mathbf{T} & \boldsymbol{\xi} \\ \mathbf{0} & 0 \end{pmatrix}, \tag{1}$$

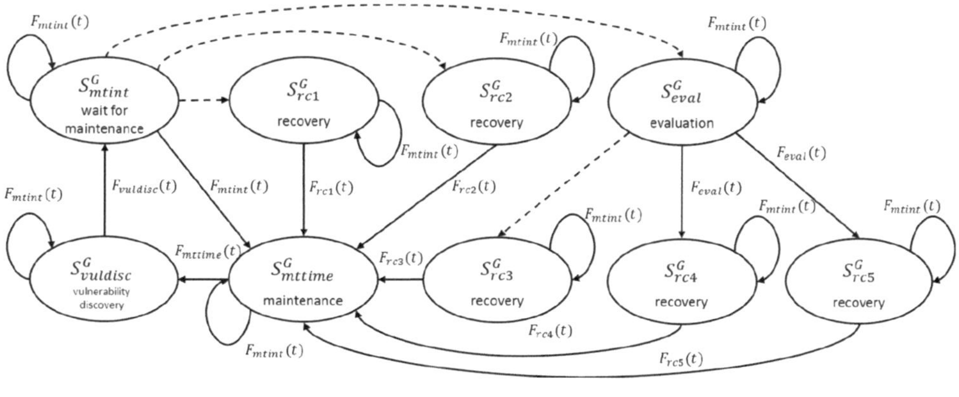

Figure 4.
State transition diagram of ITS with periodic vulnerability checking strategy.

[3] https://github.com/okamumu/JSPetriNet

where $\boldsymbol{T}$ and $\boldsymbol{\xi}$ correspond to transition rates among transient states and the exit rates from transient states to the absorbing state, respectively. Let $\boldsymbol{\alpha}$ be an initial probability vector over the transient states. Then, the cumulative distribution function (c.d.f.) of a PH-distributed variable with representation $(\boldsymbol{\alpha}, \boldsymbol{T})$ and its associated probability density function (p.d.f.) are represented by

$$F_{PH}(t) = 1 - \boldsymbol{\alpha} \exp(\boldsymbol{T}t)\mathbf{1}, \quad f_{PH}(t) = \boldsymbol{\alpha} \exp(\boldsymbol{T}t)\boldsymbol{\xi}, \tag{2}$$

where $\mathbf{1}$ is a column vector whose elements are all 1. Note that the transient states are called *phases*, and the exit rate vector is given by $\boldsymbol{\xi} = -\boldsymbol{T}\mathbf{1}$, according to the property of CTMC. In particular, the accuracy of approximation depends on the number of phases.

In the MRGP shown as in **Figure 4**, the state space is divided into nine classes (more details on MRGP state classification is referred to [18]);

- $\mathcal{S}^G_{mtint}$, consisting of the states where only GEN transition, $T_{mtinterval}$ is enabled.
- $\mathcal{S}^G_{rc1}$, consisting of the states where both GEN transitions, T_{rc1} and $T_{mtinterval}$, are enabled.
- $\mathcal{S}^G_{rc2}$, consisting of the states where both GEN transitions, T_{rc2} and $T_{mtinterval}$, are enabled.
- $\mathcal{S}^G_{rc3}$, consisting of the states where both GEN transitions, T_{rc3} and $T_{mtinterval}$, are enabled.
- $\mathcal{S}^G_{rc4}$, consisting of the states where both GEN transitions, T_{rc4} and $T_{mtinterval}$, are enabled.
- $\mathcal{S}^G_{rc5}$, consisting of the states where both GEN transitions, T_{rc5} and $T_{mtinterval}$, are enabled.
- $\mathcal{S}^G_{eval}$, consisting of the states where both GEN transitions, T_{eval} and $T_{mtinterval}$, are enabled.
- $\mathcal{S}^G_{mttime}$, consisting of the states where both GEN transitions, T_{mttime} and $T_{mtinterval}$, are enabled.
- $\mathcal{S}^G_{vuldisc}$, consisting of the states where both GEN transitions, $T_{vuldisc}$ and $T_{mtinterval}$, are enabled.

The general distributions of GEN transitions, T_x, $x \in \{mtint, rc1, rc2, rc3, rc4, rc5, eval, mttime, vuldisc\}$ are given by $F_x(t)$. In particular, we denote t_0 as the length of one checking interval, following the constant distribution:

$$F_{mtint}(t) = \begin{cases} 0 & t < t_0, \\ 1 & t \geq t_0. \end{cases} \tag{3}$$

That means, the checking interval t_0 is deterministic.

In this chapter, the general distributions are approximated by the following PH distributions:

$$\begin{aligned}
&F_{rc1}(t) \approx 1 - \alpha_1 \exp(\boldsymbol{T}_1 t)\mathbf{1}_1, \quad f_{rc1}(t) \approx \alpha_1 \exp(\boldsymbol{T}_1 t)\boldsymbol{\xi}_1,\\
&F_{rc2}(t) \approx 1 - \alpha_2 \exp(\boldsymbol{T}_2 t)\mathbf{1}_2, \quad f_{rc2}(t) \approx \alpha_2 \exp(\boldsymbol{T}_2 t)\boldsymbol{\xi}_2,\\
&F_{rc3}(t) \approx 1 - \alpha_3 \exp(\boldsymbol{T}_3 t)\mathbf{1}_3, \quad f_{rc3}(t) \approx \alpha_3 \exp(\boldsymbol{T}_3 t)\boldsymbol{\xi}_3,\\
&F_{rc4}(t) \approx 1 - \alpha_4 \exp(\boldsymbol{T}_4 t)\mathbf{1}_4, \quad f_{rc4}(t) \approx \alpha_4 \exp(\boldsymbol{T}_4 t)\boldsymbol{\xi}_4,\\
&F_{rc5}(t) \approx 1 - \alpha_5 \exp(\boldsymbol{T}_5 t)\mathbf{1}_5, \quad f_{rc5}(t) \approx \alpha_5 \exp(\boldsymbol{T}_5 t)\boldsymbol{\xi}_5,\\
&F_{eval}(t) \approx 1 - \alpha_e \exp(\boldsymbol{T}_e t)\mathbf{1}_e, \quad f_{eval}(t) \approx \alpha_e \exp(\boldsymbol{T}_e t)\boldsymbol{\xi}_e,\\
&F_{mttime}(t) \approx 1 - \alpha_m \exp(\boldsymbol{T}_m t)\mathbf{1}_m, \quad f_{mttime}(t) \approx \alpha_m \exp(\boldsymbol{T}_m t)\boldsymbol{\xi}_m,\\
&F_{vuldisc}(t) \approx 1 - \alpha_v \exp(\boldsymbol{T}_v t)\mathbf{1}_v, \quad f_{vuldisc}(t) \approx \alpha_v \exp(\boldsymbol{T}_v t)\boldsymbol{\xi}_v,
\end{aligned} \tag{4}$$

where $\mathbf{1}_1$, $\mathbf{1}_2$, $\mathbf{1}_3$, $\mathbf{1}_4$, $\mathbf{1}_5$, $\mathbf{1}_e$, $\mathbf{1}_t$, and $\mathbf{1}_v$ are the 1's column vectors, and

$$\begin{aligned}
&\boldsymbol{\xi}_1 = -\boldsymbol{T}_1\mathbf{1}_1, \quad \boldsymbol{\xi}_2 = -\boldsymbol{T}_2\mathbf{1}_2, \quad \boldsymbol{\xi}_3 = -\boldsymbol{T}_3\mathbf{1}_3,\\
&\boldsymbol{\xi}_4 = -\boldsymbol{T}_4\mathbf{1}_4, \quad \boldsymbol{\xi}_5 = -\boldsymbol{T}_5\mathbf{1}_5, \quad \boldsymbol{\xi}_e = -\boldsymbol{T}_e\mathbf{1}_e,\\
&\boldsymbol{\xi}_m = -\boldsymbol{T}_m\mathbf{1}_m, \quad \boldsymbol{\xi}_v = -\boldsymbol{T}_v\mathbf{1}_v.
\end{aligned} \tag{5}$$

Let $\boldsymbol{Q}_{x,x}$, $x \in \{\ (i)\ mtint,\ (1)\ rc1,\ (2)\ rc2,\ (3)\ rc3,\ (4)\ rc4,\ (5)\ rc5,\ (e)\ eval,\ (m)\ mttime,\ (v)\ vuldisc\}$ be the infinitesimal generator matrix of non-regenerative transitions of $\mathcal{S}_x^G$. The CTMC transition rate matrix from $\mathcal{S}_x^G$ to $\mathcal{S}_y^G$ is denoted by $\boldsymbol{Q}_{x,y}$. On the other hand, $\boldsymbol{A}_{x,y}^k$ denote the regenerative transitions from $\mathcal{S}_x^G$ to $\mathcal{S}_y^G$ triggered by transition T_k with probability $F_k(t)$, $k \in \{mtint, rc1, rc2, rc3, rc4, rc5, eval, mttime, vuldisc\}$.

Then by taking account of one checking interval t_0, the MRGP process during this interval can be approximated by the CTMC with the following infinitesimal generator as in Eq. (6), in which $\otimes$ and $\oplus$ are Kronecker product and sum. Apparently, the transition probability triggered by transition $T_{mtinterval}$ in **Figure 3c** with probability $F_{mtint}(t)$ is given by Eq. (7). In this equation, $\boldsymbol{I}$ is an identity matrix.

$$\boldsymbol{Q} = \begin{pmatrix}
\boldsymbol{Q}_{i,i} & \boldsymbol{Q}_{i,1}\otimes\boldsymbol{\alpha}_1 & \boldsymbol{Q}_{i,2}\otimes\boldsymbol{\alpha}_2 & & & & \boldsymbol{Q}_{i,e}\otimes\boldsymbol{\alpha}_e & & \\
 & \boldsymbol{Q}_{1,1}\oplus\boldsymbol{T}_1 & & & & & & & \boldsymbol{A}_{1,m}^{rc1}\otimes(\boldsymbol{\xi}_1\alpha_m)\\
 & & \boldsymbol{Q}_{2,2}\oplus\boldsymbol{T}_2 & & & & & & \boldsymbol{A}_{2,m}^{rc2}\otimes(\boldsymbol{\xi}_2\alpha_m)\\
 & & & \boldsymbol{Q}_{3,3}\oplus\boldsymbol{T}_3 & & & & & \boldsymbol{A}_{3,m}^{rc3}\otimes(\boldsymbol{\xi}_3\alpha_m)\\
 & & & & \boldsymbol{Q}_{4,4}\oplus\boldsymbol{T}_4 & & & & \boldsymbol{A}_{4,m}^{rc4}\otimes(\boldsymbol{\xi}_4\alpha_m)\\
 & & & & & \boldsymbol{Q}_{5,5}\oplus\boldsymbol{T}_5 & & & \boldsymbol{A}_{5,m}^{rc5}\otimes(\boldsymbol{\xi}_5\alpha_m)\\
 & & & \boldsymbol{Q}_{e,3}\otimes(\mathbf{1}_e\boldsymbol{\alpha}_3) & \boldsymbol{A}_{e,4}^{eval}\otimes(\boldsymbol{\xi}_e\boldsymbol{\alpha}_4) & \boldsymbol{A}_{e,5}^{eval}\otimes(\boldsymbol{\xi}_e\boldsymbol{\alpha}_5) & \boldsymbol{Q}_{e,e}\oplus\boldsymbol{T}_e & & \\
\boldsymbol{A}_{v,i}^{vuldisc}\otimes\boldsymbol{\xi}_v & & & & & & & \boldsymbol{Q}_{v,v}\oplus\boldsymbol{T}_v & \\
 & & & & & & & \boldsymbol{A}_{m,v}^{mttime}\otimes(\boldsymbol{\xi}_m\alpha_v) & \boldsymbol{Q}_{m,m}\oplus\boldsymbol{T}_m
\end{pmatrix}. \tag{6}$$

$$\boldsymbol{P} = \begin{pmatrix}
\boldsymbol{A}_{i,i}^{mtint} & & & & & & & & \boldsymbol{A}_{i,m}^{mtint}\otimes\boldsymbol{\alpha}_m\\
 & \boldsymbol{A}_{1,1}^{mtint}\otimes\boldsymbol{I} & & & & & & & \\
 & & \boldsymbol{A}_{2,2}^{mtint}\otimes\boldsymbol{I} & & & & & & \\
 & & & \boldsymbol{A}_{3,3}^{mtint}\otimes\boldsymbol{I} & & & & & \\
 & & & & \boldsymbol{A}_{4,4}^{mtint}\otimes\boldsymbol{I} & & & & \\
 & & & & & \boldsymbol{A}_{5,5}^{mtint}\otimes\boldsymbol{I} & & & \\
 & & & & & & \boldsymbol{A}_{e,e}^{mtint}\otimes\boldsymbol{I} & & \\
 & & & & & & & \boldsymbol{A}_{v,v}^{mtint}\otimes\boldsymbol{I} & \\
 & & & & & & & & \boldsymbol{A}_{m,m}^{mtint}\otimes\boldsymbol{I}
\end{pmatrix}. \tag{7}$$

We next consider the checking point when the transition $T_{mtinterval}$ fires with the probability $F_{mtint}(t)$, then the underlying process is actually an EMC with only one subspace that consists of the states where only GEN transition $T_{mtinterval}$ is enabled. Thus, the transition matrix on this regeneration point regarding $F_{mtint}(t)$ is given by

$$\boldsymbol{P}^{EMC} = \exp(\boldsymbol{Q}t_0)\boldsymbol{P}. \tag{8}$$

3.2 Availability measures

It is well known that availability is an important metric commonly used to assess the performance of repairable systems by considering both the reliability and maintainability properties of computer systems. There exist many classifications and definitions of availability, and they are used for different system environments properly. For example, when the system has a long lifetime, the steady-state availability [22] is appropriate to represent the system performance. On the other hand, when one wishes to ensure the system performance for a specific time period, the interval availability [23, 24] may be chosen to present the proportion of time during a mission or time period that the system is available for use. In this chapter, we focus on two availability criteria: interval availability and steady-state availability of the system. The interval availability is defined as the expected fraction of a given interval of time that the system is operational and is appropriate when one wishes to ensure the system availability for a specific time period. On the other hand, the steady-state availability is the limiting availability and is appropriate when the targeted system is continuously operated for a long time.

3.2.1 Interval availability

Let $\boldsymbol{\pi}_0$ denote the initial probability vector of the PH-expanded CTMC. Without loss of generality, it is assumed that the system starts at time $t = 0$. For the time interval $(0, nt_0]$, the interval availability is given by

$$\begin{aligned} A_{in}^{(n)} = \frac{1}{nt_0}(&\boldsymbol{\pi}_0 + \boldsymbol{\pi}_0\boldsymbol{P}^{EMC} + \boldsymbol{\pi}_0\boldsymbol{P}^{EMC2} + \\ &\cdots + \boldsymbol{\pi}_0\boldsymbol{P}^{EMC(n-1)}) \int_0^{t_0} \exp(\boldsymbol{Q}s)ds\boldsymbol{r}. \end{aligned} \tag{9}$$

In the above equation, $\boldsymbol{r}$ is the reward vector of the PH-expanded CTMC, and defined as

$$\boldsymbol{r} = \begin{pmatrix} \boldsymbol{r}_{mtint} \\ \boldsymbol{r}_{rc1} \otimes \mathbf{1}_1 \\ \boldsymbol{r}_{rc2} \otimes \mathbf{1}_2 \\ \boldsymbol{r}_{rc3} \otimes \mathbf{1}_3 \\ \boldsymbol{r}_{rc4} \otimes \mathbf{1}_4 \\ \boldsymbol{r}_{rc5} \otimes \mathbf{1}_5 \\ \boldsymbol{r}_{eval} \otimes \mathbf{1}_e \\ \boldsymbol{r}_{vuldisc} \otimes \mathbf{1}_v \\ \boldsymbol{r}_{mttime} \otimes \mathbf{1}_m \end{pmatrix}, \tag{10}$$

where $\boldsymbol{r}_i$ is the reward vector of system states belonging to corresponding subspace. For example, the interval availability within the first checking interval becomes

$$A_{in}^{(1)} = \frac{1}{t_0}\boldsymbol{\pi}_0 \int_0^{t_0} \exp\left(Qs\right) ds \boldsymbol{r}. \tag{11}$$

3.2.2 Steady-state availability

Using Eq. (8), the steady-state probability distribution $\boldsymbol{\pi}^{EMC} = \left(\boldsymbol{\pi}_{mtint}^{EMC}, \boldsymbol{\pi}_{rc1}^{EMC}, \boldsymbol{\pi}_{rc2}^{EMC}, \boldsymbol{\pi}_{rc3}^{EMC}, \boldsymbol{\pi}_{rc4}^{EMC}, \boldsymbol{\pi}_{rc5}^{EMC}, \boldsymbol{\pi}_{eval}^{EMC}, \boldsymbol{\pi}_{vuldisc}^{EMC}, \boldsymbol{\pi}_{mttime}^{EMC}\right)$ can be computed by solving the following linear equation:

$$\boldsymbol{\pi}^{EMC} = \boldsymbol{\pi}^{EMC}\mathbf{P}^{EMC}, \quad \boldsymbol{\pi}^{EMC}\mathbf{1} = 1, \tag{12}$$

where $\mathbf{1}$ is a column vector whose elements are 1.

Finally, we obtain the steady-state availability of the system:

$$A_{ss} = \boldsymbol{\pi}^{EMC}\boldsymbol{r}. \tag{13}$$

4. Numerical experiments

This section evaluates the interval availability and steady-state availability of the system, where the system undergoes the pull-type patch management with a periodic vulnerability checking strategy. **Table 3** gives the parameters for EXP transitions in

Parameter	Description	Value [hrs.]
$1/T_{atk}.rate$	Mean time to complete an intrusion	1200
$1/T_{undet}.rate$	Mean time passed since detection start and detection failure	8
$1/T_{det}.rate$	Mean time to detect an intrusion	12
$1/T_{fail}.rate$	Mean time to failure of a triage	6

Table 3.
Model parameters.

Notation	Transition	Distribution	Mean [hrs.]	CV
$F_{vuldisc}(t)$	$S_{vuldisc}^{G}$ to S_{mtint}^{G}	Weibull	1440	0.5
$F_{rc1}(t)$	S_{rc1}^{G} to S_{mttime}^{G}	Lognormal	24	0.5
F_{rc2}	S_{rc2}^{G} to S_{mttime}^{G}	Lognormal	12	0.5
$F_{rc3}(t)$	S_{rc3}^{G} to S_{mttime}^{G}	Lognormal	48	0.5
$F_{rc4}(t)$	S_{rc4}^{G} to S_{mttime}^{G}	Lognormal	30	0.5
$F_{rc5}(t)$	S_{rc5}^{G} to S_{mttime}^{G}	Lognormal	40	0.5
$F_{eval}(t)$	S_{eval}^{G} to S_{rc4}^{G} (S_{rc5}^{G})	Lognormal	8	0.5
$F_{mttime}(t)$	S_{mttime}^{G} to S_{mtint}^{G}	Lognormal	10	0.5

Table 4.
Probability distributions of GEN transitions.

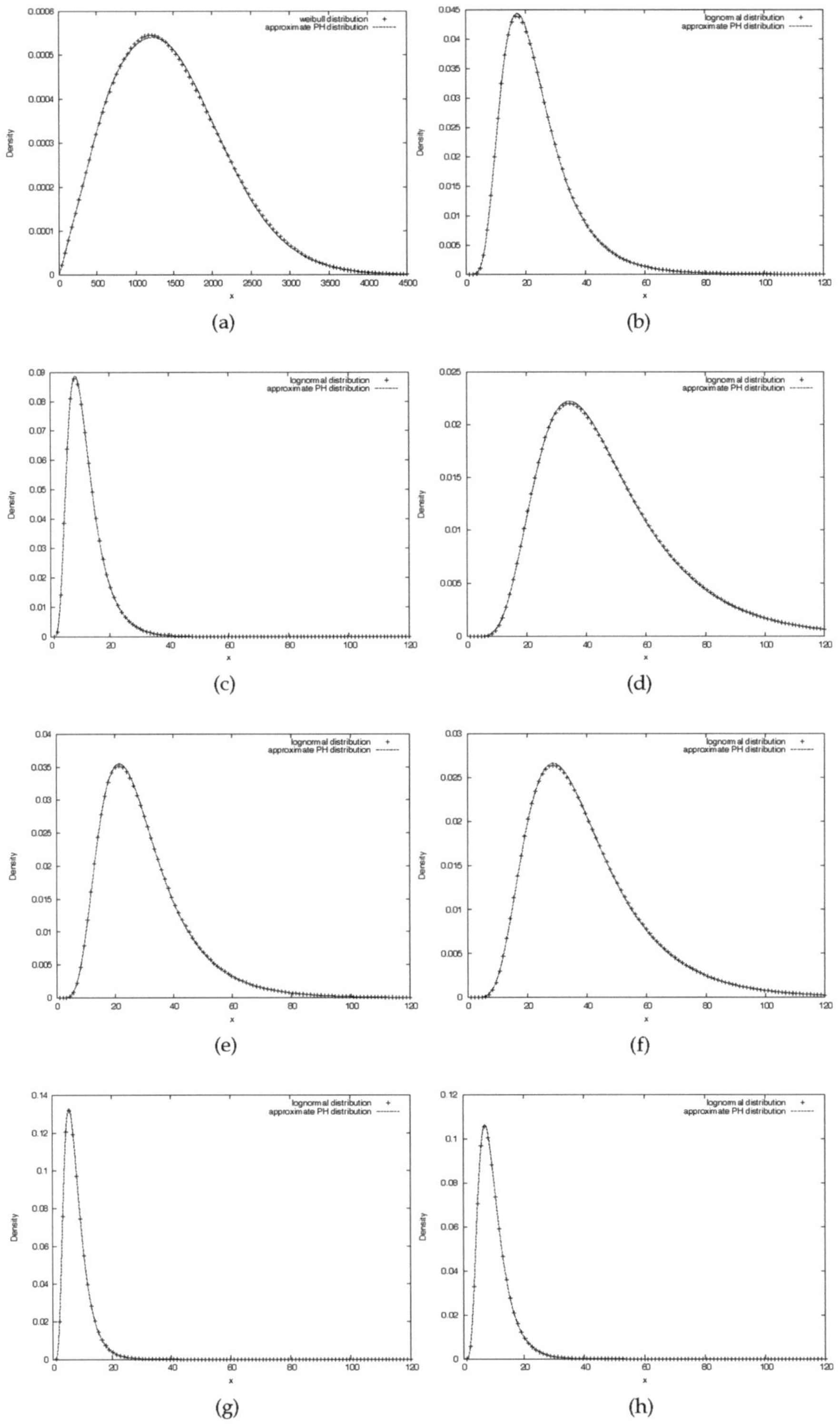

Figure 5.
Approximate PH distributions ((a) $F_{vuldisc}(t)$, (b) $F_{rc1}(t)$, (c) $F_{rc2}(t)$, (d) $F_{rc3}(t)$, (e) $F_{rc4}(t)$, (f) $F_{rc5}(t)$, (g) $F_{eval}(t)$, (h) $F_{mttime}(t)$).

Figure 3. The probability distributions for GEN transitions $T_{vuldisc}$, T_{rc1}, T_{rc2}, T_{rc3}, T_{rc4}, T_{rc5}, T_{eval}, and T_{mttime} are given in **Table 4**, where the columns of "Mean" and CV represent the mean time and the coefficient of variation, respectively.

Figure 5a–h draw the original probability distributions for GEN transitions and the approximate PH distributions with 10 phases. These figures indicate that the PH distributions are accurate enough to approximate the general distributions.

To investigate the effect of the number of checking, we consider the number of checking during 1 year, N, varying from 4 to 36. For example, in the case of $N = 4$, the length of one checking interval is 3 months. In the case of $N = 36$, the length of one checking interval is about 10 days.

Figure 6 depicts the interval availability of the system, which increases monotonically as the number of checking, N, increases. In particular, the interval availability increases sharply when the number of checking is very small. In such a case, the length of one checking interval decreased remarkably; for example, when $N = 4$, it

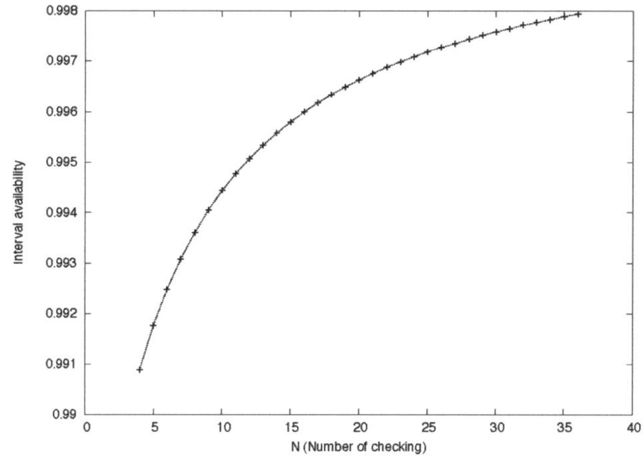

Figure 6.
Sensitivity of the number of checking on the interval availability.

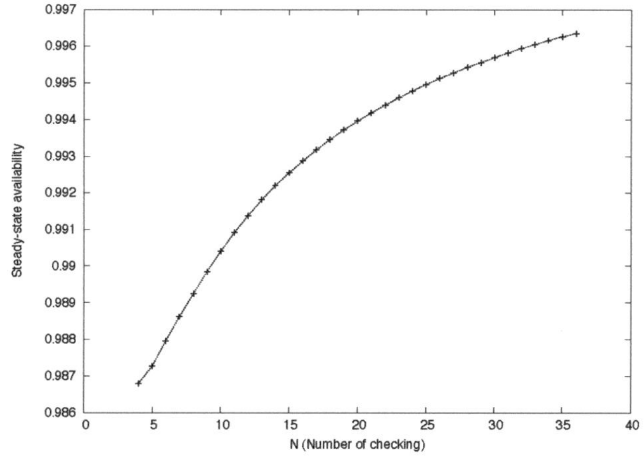

Figure 7.
Sensitivity of the number of checking on the steady-state availability.

N	t_0 [days]	Interval availability	Steady-state availability
4	91.3	0.99088	0.98679
5	73.0	0.99177	0.98726
6	60.8	0.99248	0.98795
7	52.1	0.99308	0.98862
8	45.6	0.99360	0.98925
9	40.6	0.99405	0.98985
10	36.5	0.99444	0.99040
11	33.2	0.99478	0.99091
12	30.4	0.99507	0.99137
13	28.1	0.99534	0.99180
14	26.1	0.99558	0.99219
15	24.3	0.99580	0.99254
16	22.8	0.99600	0.99287
17	21.5	0.99618	0.99317
18	20.3	0.99634	0.99345
19	19.2	0.99649	0.99372
20	18.3	0.99663	0.99396
21	17.4	0.99676	0.99418
22	16.6	0.99688	0.99439
23	15.9	0.99699	0.99459
24	15.2	0.99709	0.99478
25	14.6	0.99719	0.99495
26	14.0	0.99728	0.99512
27	13.5	0.99736	0.99527
28	13.0	0.99744	0.99542
29	12.6	0.99752	0.99555
30	12.2	0.99759	0.99569
31	11.8	0.99765	0.99581
32	11.4	0.99772	0.99593
33	11.1	0.99777	0.99604
34	10.7	0.99783	0.99615
35	10.4	0.99789	0.99625
36	10.1	0.99794	0.99634

Table 5.
The number of checking per year and its corresponding length of checking interval and availabilities.

takes almost 3 months to execute a checking operation, whereas the checking interval reduces to 2.4 months in the case of $N = 5$. However, when N increases from 35 to 36, the checking interval almost does not change. Besides, it is intuitively obvious that a shorter checking interval generally brings higher availability. Therefore, when N is a small value, the interval availability is very sensitive to the change in the value of N.

On the other hand, the steady-state availability of the system is shown in **Figure 7**. From this figure, it is found that the steady-state availability also increases as the number of checking, N, increases. Furthermore, more details on the experimental results about the number of checking per year and its corresponding length of one checking interval and availabilities are referred to **Table 5**. From this table, we can see that for any given N, the interval availability is higher than the steady-state availability.

5. Conclusion and future work

In this chapter, we presented a stochastic model to evaluate the system availability of an ITS, where the system undergoes the patch management with a periodic vulnerability checking strategy; that is, pull-type patch management. Concretely, a composite SRN model was developed to capture the overall system behaviors, including vulnerability discovery, intrusion tolerance, and reactive maintenance. Two kinds of availability criteria, the interval and steady-state availabilities, were formulated by using phase expansion. In numerical experiments, we evaluated the effect of the checking number on the system availability, and the results imply that when the checking number is small (a long checking interval), the variation in the checking number brings an significant effect into the interval availability. In addition, both interval availability and steady-state availability increase monotonically as the number of checking increases. We have also validated the accuracy of the PH approximation with 10 phases.

The chapter aims is to present a method for formulating the system availability from both transient and stationary points of view and evaluate the effect of the number of checking on the system availability. Nevertheless, it is actually well known that one of the main issues in the design of security patch management is to determine the optimal length of checking interval bringing the optimal trade-off between system performance and checking cost. For example, if the checking interval is too short, the system availability will be high, but the total checking cost will be very high. On the other hand, if the checking interval is too long, a discovered vulnerability may be exploited by malicious attacks, which decreases the system availability; in this case, the checking cost can be reduced, but the total cost due to security failures will be high. Therefore, it will be interesting, as one of future directions, to find the optimal checking number (i.e., optimal checking policy) by the consideration of both system performance and maintenance cost.

Acknowledgements

This chapter is an extension of work originally reported at the 2018 42nd IEEE International Conference on Computer Software & Applications (COMPSAC'18) [25]. Moreover, this work was supported by JSPS KAKENHI Grant Number 21 K17742.

Conflict of interest

The authors declare no conflict of interest.

Nomenclature

ITS	Intrusion tolerant system
SRN	Stochastic reward net
PH	Phase-type
MRGP	Markov regenerative process
CIA	Confidentiality, integrity, and availability
SITAR	Scalable intrusion tolerant architecture
COTS	Commercial-off-the-shelf
VM	Virtual machine
DoS	Denial-of-service
EMC	Embedded Markov chain
CTMC	Continuous-time Markov chain
GEN	Generally distributed
EXP	Exponentially distributed
c.d.f.	Cumulative distribution function
p.d.f.	Probability density function
CV	Coefficient of variation

Author details

Junjun Zheng[1]*, Hiroyuki Okamura[2] and Tadashi Dohi[2]

1 Department of Information Science and Engineering, Ritsumeikan University, Kusatsu, Japan

2 Graduate School of Advanced Science and Engineering, Hiroshima University, Hiroshima, Japan

*Address all correspondence to: jzheng@fc.ritsumei.ac.jp

References

[1] Arora A, Krishnan R, Telang R, Yang Y. An empirical analysis of software vendors' patch release behavior: Impact of vulnerability disclosure. Information Systems Research. 2010;**21**(1):115-132

[2] Abunadi I, Alenezi M. An empirical investigation of security vulnerabilities with web applications. Journal of Universal Computer Science. 2016; **22**(4):537-551

[3] Khan YI, Al-Shaer E, Rauf U. Cyber resilience-by-construction: Modeling, measuring & verifying. In: Proceedings of 2015 Workshop on Automated Decision Making for Active Cyber Defense. Denver, Colorado, USA: ACM; 2015. pp. 9-14

[4] Jansen W. Directions in Security Metrics Research. Darby, PA, USA: DIANE Publishing Co; 2010

[5] Mukkamala S, Janoski G, Sung A. Intrusion detection using neural networks and support vector machines. In: Proceedings of 2002 International Joint Conference on Neural Networks (IJCNN'02). Honolulu, HI, USA; 2002. pp. 1702-1707

[6] Stavridou V, Dutertre B, Riemenschneider RA, Saidi H. Intrusion tolerant software architectures. In: Proceedings of Darpa Information Survivability Conference and Exposition (DISCEX II'01). Anaheim, California, USA: IEEE; 2001. pp. 230-241

[7] Wang F, Gong F, Sargor C, Goševa-Popstojanova K, Trivedi KS, Jou F. SITAR: A scalable intrusion-tolerant architecture for distributed services. In: Proceedings of the 2nd Annual IEEE Systems, Man and Cybernetics Information Assurance Workshop (SMC-IAW'01). New York, USA: IEEE; 2001

[8] Zhao W. BFT-WS: A Byzantine fault tolerance framework for web services. In: Proceeding of the 11th International IEEE EDOC Conference Workshop (EDOC'07). Annapolis, MD, USA: IEEE; 2007. pp. 89-96

[9] Junior VS, Lung LC, Correia M, Fraga JDS, Lau J. Intrusion tolerant services through virtualization: A shared memory approach. In: Proceedings of the 24th IEEE International Conference on Advanced Information Networking and Applications (AINA'10). Perth, Australia: IEEE; 2010. pp. 768-774

[10] Lau J, Barreto L, Fraga JDS. An infrastructure based in virtualization for intrusion tolerant services. In: Proceedings of the 19th IEEE International Conference on Web Services (ICWS'12). Honolulu, HI, USA: IEEE; 2012. pp. 170-177

[11] Zheng J, Okamura H, Dohi T. Survivability analysis of VM-based Intrusion tolerant systems. IEICE Transactions on Information and Systems. 2015;**E-98**(12):2082-2090

[12] Kansal Y, Kapur PK, Kumar D. Assessing optimal patch release time for vulnerable software systems. In: Proceedings of 2016 International Conference on Innovation and Challenges in Cyber Security. Greater Noida, India: IEEE; 2016. pp. 308-314

[13] Uemura T, Dohi T, Kaio N. Availability analysis of an Intrusion tolerant distributed server system with preventive maintenance. IEEE Transactions on Reliability. 2010;**59**(1): 18-29

[14] Wang D, Madan BB, Trivedi KS. Security Analysis of SITAR intrusion tolerance system. In: Proceedings of the

2003 ACM Workshop Survivable and Self-regenerative Systems: in association with 10th ACM Conference on Computer and Communications Security (SSRS'03). Fairfax, VA, USA: ACM; 2003. pp. 23-32

[15] Çinlar E. Introduction to Stochastic Processes. Englewood Cliffs, NJ, USA: Prentice-Hall Inc; 1975

[16] Fricks R, Telek M, Puliafito A, Trivedi KS. Markov renewal theory applied to performability evaluation. In: Bagchi K, Zobrist G, editors. State-of-the Art in Performance Modeling and Simulation. Modeling and Simulation of Advanced Computer Systems: Applications and Systems. Amsterdam, The Netherlands: Gordon and Breach Publishers; 1998. pp. 193-236

[17] Garg S, Pfening S, Puliafito A, Telek M, Trivedi KS. Analysis of preventive maintenance in transaction based software systems. IEEE Transactions on Computers. 1998;**47**(1): 96-107

[18] Zheng J, Okamura H, Li L, Dohi T. A comprehensive evaluation of software rejuvenation policies for transaction systems with MarMarkov arrival. IEEE Transactions on Reliability. 2017;**66**(4): 1157-1177

[19] Okamura H, Yamamoto K, Dohi T. Transient analysis of software rejuvenation policies in virtualized system: phase-type expansion approach. Quality Technology & Quantitative Management. 2014;**11**(3):335-351

[20] Okamura H, Dohi T. A phase expansion approach for transient analysis of software rejuvenation model. In: Proceedings of the 8th International Workshop on Software Aging and Rejuvenation (WoSAR'16). Ottawa, Canada: IEEE; 2016. pp. 98-103

[21] Choi H, Kulkarni VG, Trivedi KS. Markov regenerative stochastic Petri nets. Performance Evaluation. 1994;**20**: 337-357

[22] Hosford JE. Measures of dependability. Operations Research. 1960;**8**(1):53-64

[23] Rubino G, Sericola B. Interval availability analysis using operational periods. Performance Evaluation. 1992; **14**(3–4):257-272

[24] Smith M, Aven T, Dekker R, van der Duyn Schouten FA. A survey on the interval availability distribution of failure prone systems. In: Advances in Safety and Reliability: Proceedings of ESREL'97. Oxford: Elsevier; 1997. pp. 1727-1737

[25] Zheng J, Okamura H, Dohi T. A pull-type security patch management of an intrusion tolerant system under a periodic vulnerability checking strategy. In: Proceedings of the 2018 IEEE 42nd Annual Computer Software and Applications Conference (COMPSAC'18). Tokyo, Japan: IEEE; 2018. pp. 630-635

Chapter 6

Automated Condition Monitoring of a Cycloid Gearbox

Eric Bechhoefer

Abstract

While condition monitoring techniques have been developed for many gearbox types, there has been almost no research on condition monitoring of cycloid driver gearboxes. Cycloid gearboxes are used where high reduction ratios are needed in a single stage. While most gear designs are based on an involute subject to a sliding force, cycloid gear designs are subject to compression. As a result, cycloid gearboxes are quiet, have low backlash, and have large torsional stiffness. Because there is no typical pinion-gear pair in this gearbox, the calculation of the reduction ratio is non-standard. Further, as the eccentric bearing which drives the cycloid gears is in the rotating frame, the calculated fault frequency rates are not as expected. This paper describes the dynamics needed to identify cycloid gearbox fault features to achieve automated fault detection and alerting.

Keywords: TSA, threshold setting, bearing envelop analysis, resonance, model-based dynamics

1. Introduction

There are few non-standard condition monitoring applications for unusual gearbox designs. While most reduction (example: epicyclical) gearboxes have well-understood dynamics, others, such as a Variator (continuously variable transmission) or cycloidal gearbox, have not been reported. This paper covers the dynamics, configuration, and some test observations of work done on a cycloidal gearbox. The analysis procedure applies to other, more standard gearboxes as well.

In a cycloidal gearbox, the drive uses an eccentric bearing that causes the gears' center to rotate in a housing. The rotation orbit is reversed as the gear's teeth are less than the housing's diameter. The path of a fixed point of the gear traces a hypocycloid, which is fundamentally different from the circular motion of traditional gears.

Interest in the cycloidal gearbox is derived because they are used in many applications where low-cost drive motors are needed. For example, many conveyer belt systems (sorting, moving bulk media, slew drives) use cycloid drives. Typically, the drive itself is a relatively low-cost asset. However, the processes that drive unit support can have significant economic impacts if they fail. For example, one of the more significant courier delivery services has a distribution center with 3000 cycloid drives to move packages. The loss of a drive unit halts sorting, impacting up to $200,000 per day in revenues.

The cycloid drive for gearboxes allows for a high reduction ratio and zero or very low backlash. The cycloid gear design is based on compression vs. shear forces, where the contact is typically rolling vs. sliding. These features allow high shock load capacity, high torsional stiffness, and quiet operation. Single-stage ratios of more than 200:1 are possible.

The gearbox chosen for the test was an integrated induction motor and gearbox. This gearbox is rated for 0.75 kW, approximately 1 Hp drive. For 60 Hz operations, using a four-pole motor, the drive unit has a 100% input shaft rate of approximately 1795 rpm. The gearbox has a 51:1 reduction ratio.

2. State of the art for gearbox condition monitoring

There has been little documented work on the condition monitoring of cycloidal gearboxes. Chrochran and Bobak [1] describe the complexity of vibration analysis of cycloid gearbox using traditional spectrum analysis. They give information on calculating the cycloidal disc mesh frequency but do not describe a method for automated fault detection. There is no process given for bearing fault frequency indication.

Condition monitoring of motors or gearboxes has generally used spectrum analysis. The spectrum measures the magnitude of a frequency associated with the component fault frequencies, such as a shaft or gear mesh. The Fourier transform, used in spectrum analysis, is defined by cosines. The spectrum is good at measuring periodic sinusoids. However, many faults result in impacts that are not well measured by the spectrum. In recognizing this, R.M. Stewart [2] ushered in modern gearbox analysis using the Time Synchronous Average (TSA). The TSA, which controls for variance in speed, also performs as a comb filter that rejects nonsynchronous vibration features. The resulting time-domain signal reveals impact features. These features can be quantified via statics indicators, such as RMS, kurtosis, skewness, or crest factor [3, 4].

3. Vibration-based condition monitoring

Condition monitoring uses vibration sensors and configuration representing the drivetrain/motor to calculate condition indicators (CIs). These CIs are used to infer the current health of the component, and with the current component health, and the component threshold, an estimate the remaining useful life (RUL) can be estimated. The RUL (e.g., prognostics) allows the operator to better manage the asset by scheduling maintenance opportunistically. The goal, along with increased asset safety, is improved availability and more opportunities for revenue generation.

Some CIs have physical meaning. For example, a shaft imbalance is measured by the 1st shaft harmonic (SO1), typically as a velocity such as inches per second (IPS). ISO 10816 Vibration gives direction on these limits for various equipment types. Other fault conditions for shafts could include bending or coupling issues, which excite high harmonics. For these faults, there are no standard limits. Similarly, for components such as gears and bearings, the CIs have little physical meaning, and statistical or machine learning methods are used to set a threshold representative of a fault condition.

Acceleration, the second derivative of displacement, is a function of the shaft rate to the second power. Hence, acceleration from high-speed shaft tends to dominate simpler time-domain statics such as RMS. The vibration spectrum can give magnitude for a given shaft or gear mesh in the frequency domain. This is valid if the shaft rate is relatively stable RPM. However, for many systems, Fourier analysis, of say, the Gear mesh frequency is not necessarily a good fault indicator. For bearings, detection of the fault frequency (Cage, Ball, Inner, or Outer Race rate) is only possible close to failure. For these reasons, more advanced analyses are required.

Signal processing techniques such as the Time Synchronous Average (TSA) is used to control for variance in shaft rate and are the basis of gear condition indicators. As the TSA is a time-domain analysis, it is sensitive to impacts associated with, for example, a breathing crack. However, TSA does not work for bearing analysis. Bearings require other techniques because their motion depends on non-Hertzian contact resulting in slip (by definition, nonsynchronous). Additionally, due to the nature of bearing faults (e.g., we are measuring the effect of an impact inducing resonance in the bearing itself), successful fault detection requires careful consideration of parameter inputs necessary (e.g., envelope window) to perform the analysis.

3.1 Analysis based on the time synchronous average

Modern techniques for vibration diagnostics using the TSA were introduced in "Some Useful Data Analysis Techniques for Gearbox Diagnostics" [2]. In addition to the TSA, Stewart proposed several new gear fault condition indicators. These gear algorithms and subsequent new analyses by McFadden [3], Ma [5], and others are based on the functions operating on the TSA.

The model for vibration in a shaft in a gearbox was given in [2] as:

$$x(t) = \sum_{i=1}^{k} X_i(1 + a_i(t)) \cos\left(2\pi i f m(t) + \phi_i(t)\right) + b(t) \quad (1)$$

where:
X_i is the amplitude of the kth mesh harmonic.
$FM(t)$ is the average mesh frequency.
$a_i(t)$ is the amplitude modulation function of the ith feature harmonic.
$\phi i(t)$ is the phase modulation function of the ith feature harmonic.
$b(t)$ is additive background noise.

The mesh frequency is a function of the shaft rotational speed: $FM = Nf(t)$, where N is the number of teeth on the gear and $f(t)$ is the shaft speed as a function of time. As most drive motors are induction machines, the slip, and hence, motor speed, will change based on changes in torque over time. This will cause the resulting spectrum to be smeared in the frequency domain.

The vibration data can be resampled with a tachometer signal (such as a key phasor) and with the ratio from the key phasor to the shaft under analysis. The number of data points between one revolution and the next revolution is the same. The time-synchronous averaging (TSA), sums each point over the revolution, with the resampled data, then divides by the number of revolutions during the acquisition.

Since the radix-2 FFT is most used, the number of data points in one shaft revolution (n) is interpolated into m number of data points, such that:

- For all shaft revolutions n, m is larger than r (the number of samples in one revolution), and
- m = 2ceiling ($\log_2 (r)$)

The TSA acts as a comb filter, where the passband (comb) is each shaft harmonic. This removes nonsynchronous signals from the TSA. Operations on the TSA, such are RMS or magnitude of the first harmonics of the Fourier transform of the TSA, define various Condition Indicators (CIs). There are many potential CIs, which may include the second and third harmonics derived from the spectrum of the TSA, and other statistics such as: Peak to Peak, or kurtosis (**Figure 1**).

As there are gears associated with the input/output shaft, further analysis is performed on the TSA and the spectrum of the TSA. Some analyses are classified as gear specific, which use the number of teeth on the gear under analysis (FM0 [2], the AM/FM analysis [3], for example). Other non-gear-specific analyses are also performed, such as the residual or the energy operator (a time/frequency analysis). It should be noted that there are many implementations of gear analysis [4], and there is no single analysis that works for every gear fault type. In this implementation, the system generated 18 CIs for each gear (**Figure 2**).

3.2 Basic gear analysis

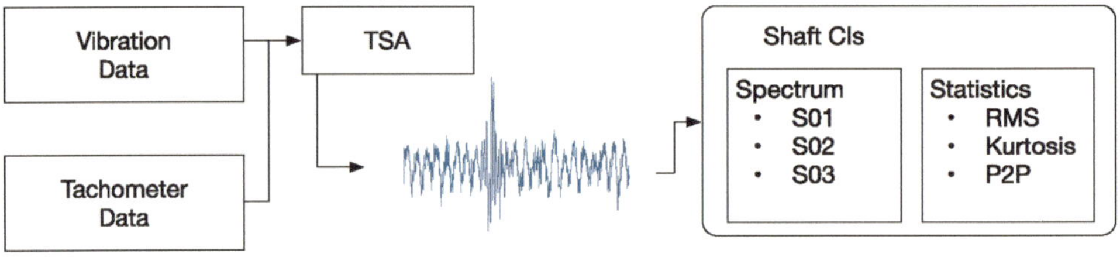

Figure 1.
Calculation of the TSA.

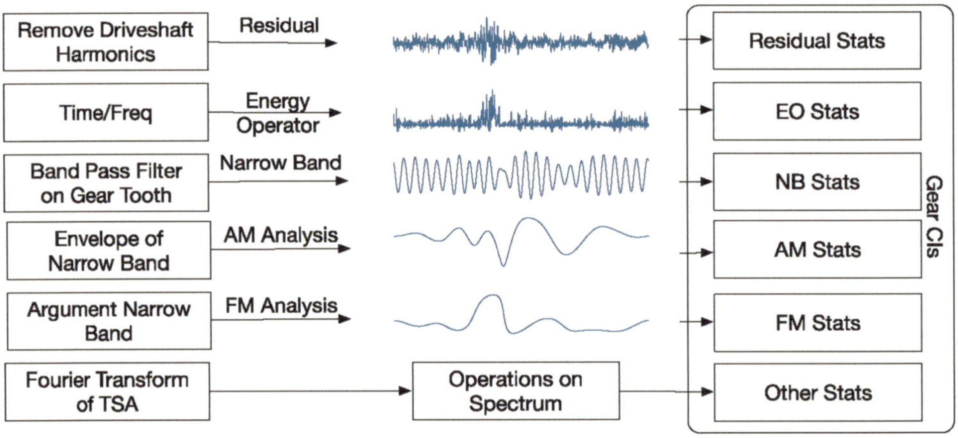

Figure 2.
Operating on the TSA to generate gear condition indicators.

In the residual gear analysis, the Fourier transform of the TSA is taken, and harmonics associated with shaft and gear mesh are zeroed, then the inverse Fourier transform is taken. In the frequency domain of the TSA, each index is a shaft harmonic. Hence if there are 31 teeth on a gear, this mean the 32 index in the frequency domain (index 1 is DC) is the 1st gear mesh harmonic. Removing this and 2nd/3rd gear mesh harmonics in the frequency domain removes these superimposed tones in the time domain. Without these known periodic signals in the time domain TSA signal, non-periodic features, such as the impact of a broken tooth, can be identified in the waveform.

Damage to a component change the measured instantaneous frequency will. The energy operator, developed by Ma [5], can quantify the amplitude and phase-modulated signal of a fault, whose product can measure the instantaneous frequency due to say, a scuff or cracked tooth. The energy operator is sensitive to torque, so statical indicators that reflect distribution "shape," such as kurtosis and crest factor, can be used. The EO is given as:

$$\Psi_{EO}(TSA_n) = TSA_n^2 - TSA_{n+1} \times TSA_{n-1} \tag{2}$$

The Narrowband analysis [3]. uses the Fourier transform as a bandpass filter to remove all frequencies not associated with the gear mesh. The selection of the filter bandwidth is usually 25% of the gear tooth count. For example, if the TSA is length 1024, and the gear tooth count is 31, the filter bandwidth is 31/4 = 8, or [23 to 39]. After taking the Fourier transform of the TSA, the from DC to index (31–8) = 23, and index 39 to 512 are set zero (along with their conjugate). The Narrowband signal is then the real part of the inverse Fourier transforms.

The AM (amplitude modulation) Analysis is the envelope of the narrowband signal. Essentially, this is simply the magnitude of the Hilbert transform. Similarly, the FM (frequency modulation) Analysis is derivative (instantaneous frequency) of the argument of the Hilbert transform.

The TSA Fourier transform is used for miscellaneous analysis [2, 4, 5]. The Figure of Merit 0, for example, is a well know analysis [2] and is generally calculated as:

$$\text{FM0} = tsapeaktopeak \Big/ \sum_{i=1}^{3} GM_i \tag{3}$$

GM_i is the ith gear mesh harmonic. As the TSA is a time-domain signal, the peak-to-peak value is the maximum of the TSA time domain value minus the minimum of this TSA time domain value. In general, the peak to peak will increase over time as if there is a propagating cracked or soft tooth, while the gear mesh harmonics will remain constant. FM0 can be a powerful indicator of a crack or soft tooth.

The residual RMS is sensitive soft/crack tooth, as the residual of the TSA does not remove features associated with the impact of a breaking crack. The RSM of the TSA will is not as sensitive to these impact events. As such, the ratio of the residual RMS to the TSA RMS can be a helpful condition indicator, defining the energy ratio. If the residual signal is defined as r, and the TSA is $tsa_{i,}$ then.

$$er = \sqrt{\frac{\sum_{i=1}^{n}(r_i - \bar{r})^2}{n}} \Big/ \sqrt{\frac{\sum_{i=1}^{n}(tsa_i - \overline{tsa})^2}{n}} \tag{4}$$

The sideband level factor is defined as the sum of the first-order sideband amplitudes about the gear mesh, divided by the TSA rms [4]:

$$SLF = TSA_{gm-1} + TSA_{gm+1} / \sqrt{\frac{\sum_{i=1}^{n} \left(tsa_i - \overline{tsa}\right)^2}{n}} \tag{5}$$

The ratio of the second gear mesh harmonic energy ratio to the first gear mesh harmonic energy defines the G2 analysis. Example analyses are found in the appendix, and a full description is given in [2–6].

3.3 Bearing envelope analysis

Bearing analysis is a separate processing flow. Bearings, as they are designed to be greased/oiled, have non-Hertzian contact. Typically, we observe a 1% slip in the calculated motion of the bearing components. Some bearings, when under thrust, will have changed their contact angle and pitch diameter, resulting in an increased fault rate by 2 to 3% [7, 8]. The bearing analysis is asynchronous but must also consider the non-stationarity of the shaft. To control for changing shaft rate, the vibration data can be resampled [8]. Bearing analysis uses this speed corrected signal for envelope analysis, which takes the spectrum of the demodulated signal and envelopes (absolute value of the Hilbert transform) the vibration data (**Figure 3**).

The bearing analysis process returns seven CIs for each bearing, including the cage, ball, inner and outer race energies, the 1/rev spectral energy, the whip/whirl energy (for journal bearing analysis), and the kurtosis of the spectrum.

3.4 Health indicator paradigm

Intending to automate fault detection, we wish to use the calculated CIs to infer the health of a component [4]. Defining a health indicator (HI) assumes that CIs have some distribution. The HI is then a function of distributions. This allows a rigorously defined threshold setting process for a given false alarm rate. With that in mind, one can define the HI such that:

- The HI is scaled from 0 to 0.35, where 0.35 is the PFA (probability of false alarm). The PFA is set to say 10e-6, which is small,
- When the HI is greater than 0.75, the component is in warning. The probability of false alarm is then minimal for a nominal component.

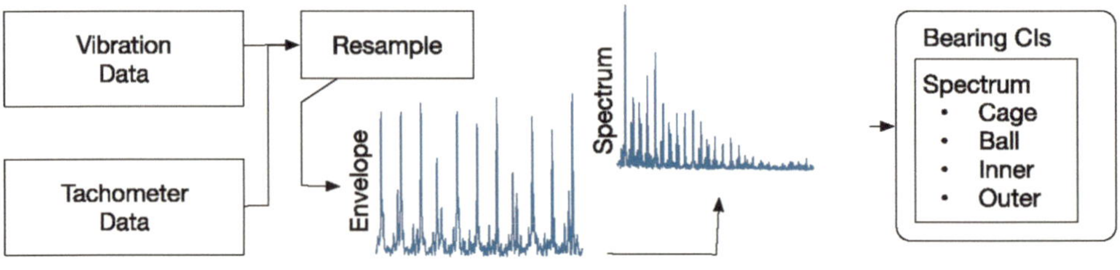

Figure 3.
Bearing analysis process flow.

- When the HI is greater than 1.0, the component is in alarm.

It is not claimed that the HI is a measure of failure. An HI based on the function of distributions develops evidence to reject the Null hypothesis: that the component is nominal. When the hypothesis is rejected, e.g., the HI is greater than 1.0, evidence suggests that the component is damaged. Hence, it allows for a proactive maintenance policy to restore the component to its nominal condition through repair. Proactive maintenance protects against cascading damage and reduces gearbox replacements.

The HI paradigm, from a maintainer perspective, is a stoplight-based threshold setting/alerting system: when a component is yellow, plan maintenance, and when the component turns red, do maintenance.

3.5 Controlling for the correlation between CIs

It is assumed that CIs have a probability distribution (PDF). Operation on the CI to build an HI is then a function of distributions. The norm of the CIs is the HI function used in this test:

$$HI = 0.35/crit\sqrt{\boldsymbol{Y}^T\boldsymbol{Y}} \tag{6}$$

where $\boldsymbol{Y}$ is the whitened, normalized array of CIs, and *crit*, is the critical value.

Only if the CIs are independent and identical (e.g., IID) is (6) valid. For Gaussian distribution, subtracting the mean and dividing by the standard deviation will give identical Z distributions. Ensuring the independence of a vector of CIs is much more difficult. In **Table 1**, the correlation coefficients for 6 CIs used for gear fault analysis: most correlation values are statically significant. Hence preprocessing is needed to whiten the CIs for (6) to be valid.

This correlation between CIs implies that for a given function of distributions to have a threshold that operationally meets the design PFA, the CIs must be whitened (e.g., de-correlated). The Cholesky decomposition was used as a whitening function, as the Cholesky decomposition of the Hermitian is always positive definite. If the inverse correlation matrix of the ***CIs*** is Σ^{-1}, then:

$$\boldsymbol{LL}^* = \Sigma^{-1}, \text{then } \boldsymbol{Y} = \boldsymbol{L}\, x\, \boldsymbol{CI}^T \tag{7}$$

Where L is a lower triangular, and L^* is its conjugate transpose. $\boldsymbol{Y}$ is 1 to n independent CI with unit variance (one CI representing the trivial case).

ρij	CI 1	CI 2	CI 3	CI 4	CI 5	CI 6
CI 1	1	0.84	0.79	0.66	−0.47	0.74
CI 2		1	0.46	0.27	−0.59	0.36
CI 3			1	0.96	−0.03	0.97
CI 4				1	0.11	0.98
CI 5					1	0.05
CI 6						1

Table 1.
Correlation coefficients for the six CIs used in the study.

3.6 Finding the critical value

The critical value is calculated by using the inverse cumulative distribution function for the HI. In this example, it was assumed that the CIs had Rayleigh PDFs, or through a simple transformation, made to approximate Rayleigh. This assumption was made because for magnitude-based CIs, it can be shown that the CI PDF is Rayleigh. In the case of Gear or Bearing CIs (where a DC offset biases magnitudes), the bias is removed to make CIs approximate Rayleigh.

The Rayleigh PDF has some nice properties. For one, Rayleigh distribution uses a single parameter, β, defining the mean $\mu = \beta^*(\pi/2)^{0.5}$, and variance $\sigma^2 = (2 - \pi/2) * \beta^2$. The PDF of the Rayleigh is: $x/\beta^2 \exp(x/2\beta^2)$. When applying these equations to the whitening process, the value for β for each CI will be: $\sigma^2 = 1$, and$\beta = \sigma^2 / (2 - \pi/2)^{0.5} = 1.5264$.

The HI derived from (6), will have a Nakagami PDF [3]. The statistics for the Nakagami are $\eta = n$, and $\omega = 1/(2-\pi/2)*2*n$, where n is the number of IID CIs used in the HI calculation.

4. The cycloid gearbox

The main components of the gearbox are the input shaft, input shaft support bearing, two eccentric bearings, the cycloid gears, the pin teeth-case, the pins, output rollers, output shaft, and the output support bearing. The ratio for the gearbox is given as:

$$\text{ratio} = (n_{teeth-1}) \times n_{pins} / (n_{teeth} - n_{pins}) \tag{8}$$

The test gearbox has a dual disc with 26 teeth and 51 pins.

4.1 Equations of motion and configuration

Configuration is driven by the equations of motions for the monitoring components. This consists of describing synchronous motion analysis of the shafts and gears and the asynchronous motion of the bearings.

The simple input/output gearbox design uses three bearings on the input shaft: bearing D (the eccentric bearing) and input shaft bearing C. Two bearings support the output shaft: bearing B and bearing A.

The shaft rate determines the bearing rate fault frequencies and:

- the number of rolling elements (b),
- the roller element diameter (d),
- the bearing pitch diameters (e), and
- the bearing contact angle (α).

The fault features are related to damage accumulated on the bearing itself.

There are typically six fault features calculated for the bearing associated with bearing elements: cage, ball, inner race, outer race. For mechanical looseness, the bearing may also generate signatures associated with whip/whorl (in the base spectrum) or a 1/revolution impact (tick) in the heterodyne analysis. The bearing feature rates are calculated as:

$$cage = 0.5(1 - d/e * \cos(\alpha)) \tag{9}$$

$$ball = e/d\left(1 - (d/e)^2 * \cos(\alpha)^2\right) \tag{10}$$

$$innerrace = b/2(1 + (d/e) * \cos(\alpha)) \tag{11}$$

$$outerrace = b/2(1 - (d/e) * \cos(\alpha)) \tag{12}$$

Because the outer rate of the eccentric bearing is in contact with the cycloid gear and the input shaft, the total rate seen by the bearing is the input shaft + output shaft. The eccentric bearing analysis was assigned to the input shaft. To capture the change in the relative motion of the bearing to the shaft, the bearing rates were corrected by 1 + 1/51 = 1.0196. This is used to determine the correct bearing rate fault features.

Shaft and gear analyses are based on the time-synchronous average, which requires an accurate ratio from the tachometer. The tachometer is used to resample the vibration data and correct for any changes in shaft rate. Gear analysis, and more importantly, gear mesh frequencies, is a function of the shaft rate and the number of teeth on the gear. In a traditional gearbox, an input shaft with a 29.23 Hz rate with 26 teeth would have a gear mesh frequency of 29.23 × 26 = 759.96 Hz. However, in the cycloid gear, the relative motion to the shaft is driven by the eccentric gear and the output shaft. The motion of the cycloid to the ring gear has, for each revolution, one extra gear mesh. The actual gear mesh frequency is then 789.19 Hz. For this reason, gear analysis is based on 27 teeth, not 26 teeth.

Normally, the ring gear analysis would usually be associated with the number of ring gears teeth. However, there are pairs of cycloid gears (of 26 teeth), resulting in a measured mesh of 51 × 2 or 102 mesh. The TSA spectrum and raw spectrum then show frequencies at 29.23/51 × 102 = 58.46 Hz. Due to the modulation of two-disc, there are sidebands at 102 +/−51 = 51 and 153, or 29.22 and 87.69 Hz.

Example CIs used for the analysis were: Residual RMS, Residual Kurtosis, Residual Crest Factor, Energy Ratio, Energy Operator Kurtosis, Energy Operator Crest Factor, Figure of Merit 0, Side Band Lifting Factor, Side Band Analysis, Narrow Band Kurtosis, Narrow Band Crest Factor, Amplitude Modulation RMS, Amplitude Modulation Kurtosis, Frequency Modulation RMS, Frequency Modulation Kurtosis, Gear Mesh Energy (reference [4], see appendix for Matlab © source code for these analyses).

The envelope analysis is based on the demodulation of high-frequency resonance from impact s(bearing envelope analysis is given in the appendix). Poor selection of a window results in poor envelope/bearing analysis. In general, techniques such as spectral kurtosis have been used to select envelop windows, but it is not easy without fault data. Alternatively, a simple calculation of the resonance can be performed.

Lord Rayleigh [9] equated kinetic energy at the mean position of a beam to strain energy at the maximum displacement on a ring with a similar nodal configuration. This can be used to estimate the resonance of a ring, such as a bearing. When evaluated, this equation seemed to underestimate the natural frequency of the bearing when tested. Timoshenko [10] further developed the concept of Rayleigh to calculate the natural frequency of a ring. Timoshenko teaches that for a ring with uniform mass, the exact shape of the mode of vibration consists of a curve which is a sinusoid on the developed circumference of the ring.

The natural frequencies are then:

$$\omega_s = n\left(n^2 - 1\right)/\sqrt{n^2 + 1}\sqrt{EI/\mu R^4} \tag{13}$$

where:
μ is the mass per unit length,
EI is the bending stiffness (Youngs Modulus × Inertia).
R is the radius.
Window selection is based on the sample rate of the sensor. The sample rate also affects the length of the TSA:

$$TSAlength = 2^{ceil\left(log_2(SampleRates/ShaftRate)\right)} \tag{14}$$

Given the low output shaft rate of approximately 0.57 Hz, the measured acceleration will be low. For this reason, the acquisition length must be adequately long to capture perhaps 20 revolutions. Hence, a high sample rate taken over an extended period results in a large data set, which takes more time to process and download raw data (if needed).

For this reason, the sample rate of the output shaft was taken at 2930 sps for 60 seconds. As the output shaft rate is 0.57 Hz, this collects 34 revolutions. The TSA length is then 8192. For the input shaft, which is closer to 30 Hz, only 8 s of data were taken at 23438. This allows a Nyquist frequency of 1465 Hz for the output shaft and 11719 for the input shaft. From the model response of Eq. (13), the window for output shaft analysis was taken at 300 to 1300 Hz, which covers the small resonant mode at 1000 Hz. The window was taken from 9 to 11 kHz for the input shaft, covering the modal response at 10 kHz.

5. Test stand results

We ran the gearbox unit at approximately 50% load for 45 hours using a nominal gearbox. Acquisitions were taken every 5 min. This allowed us to collect healthy gearbox data from which we could set thresholds as per (6). After the initial test run to set thresholds, the gearbox was run at 150% torque load for 1 h. The high torque load was used to initiate a propagating fault. The gearbox was then run 100% (rated torque) until failure (e.g., the gearbox seized due to a failure of the output bearing). In general, vibration data indicated multiple damaged components because of the torque overload.

For example, clearing seen in **Figure 4** is the step change due to the overload at time – 175 hours, followed by an increasing trend/imbalance in the input shaft. The imbalance in the input shaft was due to the eccentric bearing being damaged during 150% loading.

Surprisingly, while reflecting the damage initiation, the output bearing only began the trend to failure toward the end of the run (**Figure 5**).

The envelope spectrum of the failed output bearing 1 day prior to failure (**Figure 4**) shows mechanical looseness. The mechanical looseness is seen at the 1/Rev. at 27 Hz. Additionally, the ball rolling elements and outer race were damaged. Note that the rolling elements and outer race fault are approximately 1% below the calculated rate due to slip (**Figure 6**).

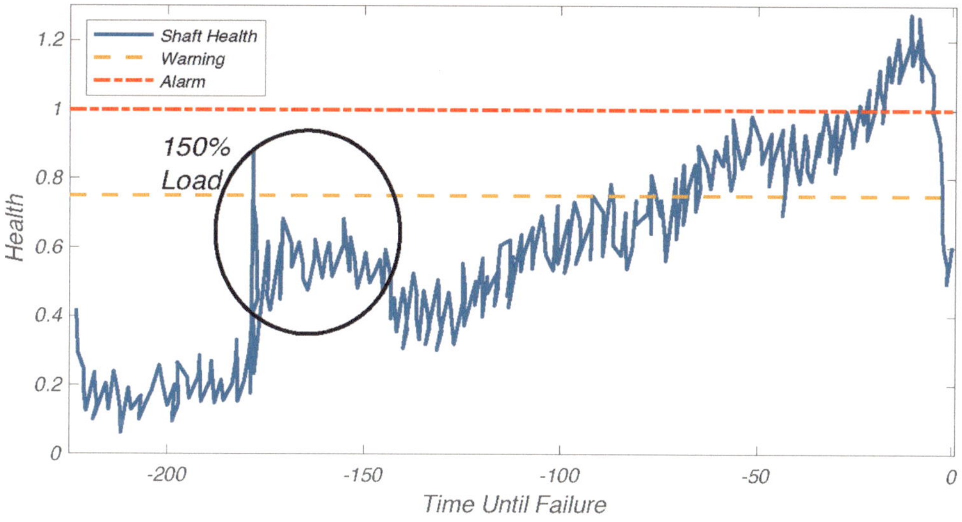

Figure 4.
Input shaft health. Step change occurs from 150% overload.

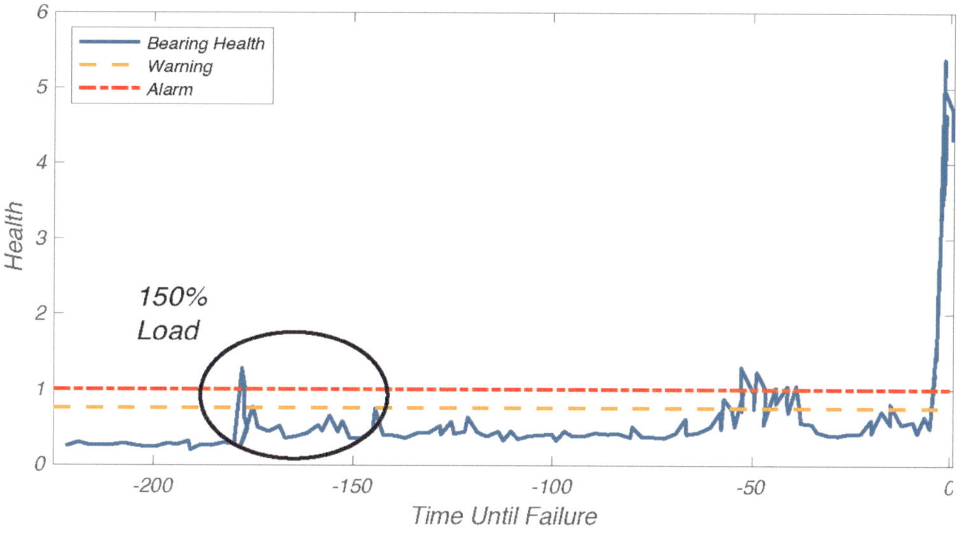

Figure 5.
Output bearing health vs. time.

Both the cycloid gear and ring gear also showed damage propagation. The cycloid gear shows in alarm level gear mesh 50 to 20 hours before failure. This was driven predominately by gear mesh energy, which is not usually a consistent indicator of damage (**Figure 7**).

Note that from 20 hours before failure, Residual RMS, Energy Ratio, and FM0 are sensitive to the impending fault. From this, it was learned that the best five indicators for the cycloid gear health: Residual RMS, Energy Ratio, FM0, AM Kurtosis, and Gear Mesh. This suggests that during the last 10 to 20 hours of the run, the cycloid gear experienced a second failure mode detected by the more traditional gear faults.

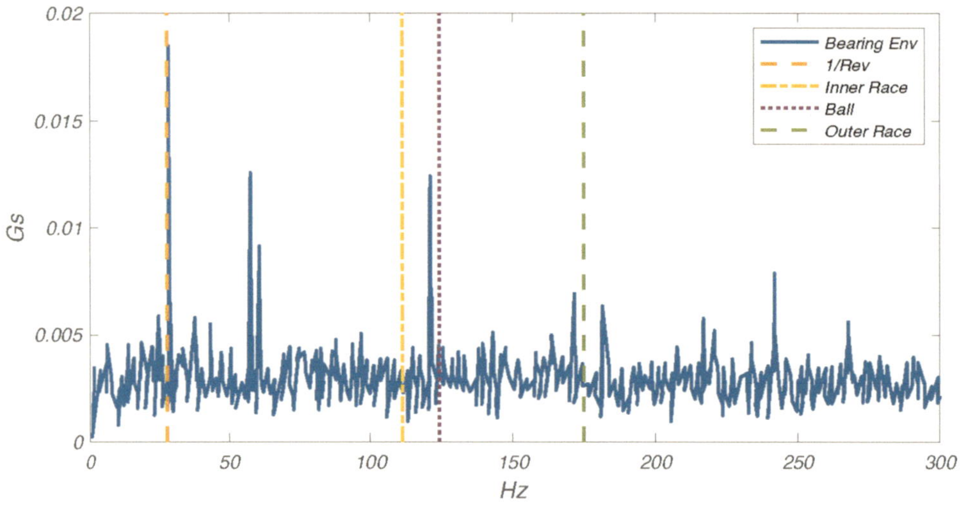

Figure 6.
Output bearing envelope Spectrum.

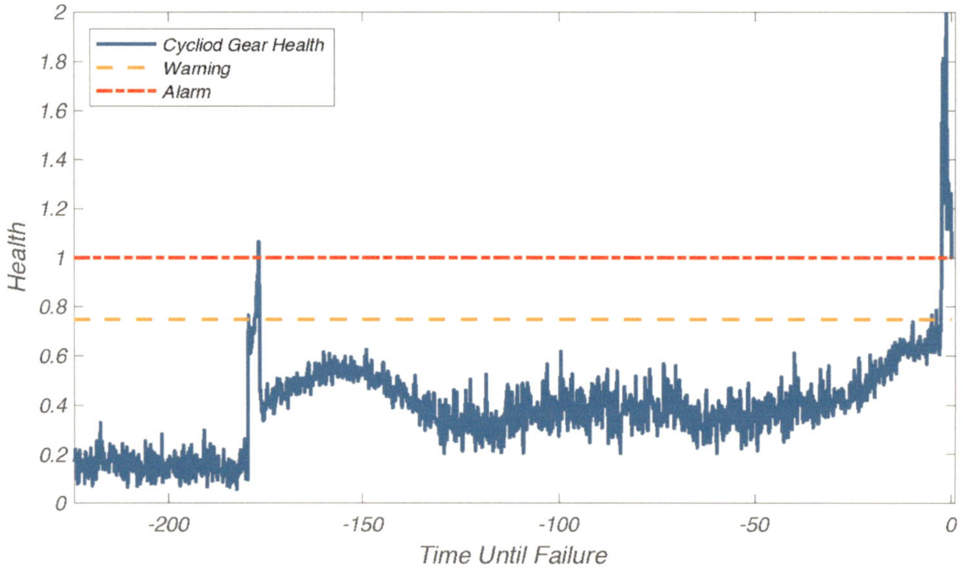

Figure 7.
Cycloid gear heath vs. time.

6. Conclusion

The cycloid gearbox has unique dynamics, requiring the correct ratios for the TSA and bearing rate calculation. The bearing analysis for the Cycloid gearbox is relatively standard. The eccentric bearing rate was multiplied by a correction factor to account for the rotating frame of the outer race. During the run to failure test, the eccentric bearing roller elements and the output bearing rolling elements were faulted. Toward the end of the run to failure test, the high level of damage (resonance energy) associated with the eccentric gear raised the noise floor of the envelope analysis, which contributed to the high HI level of all the bearings in the gearbox.

The cycloid gear itself posed a challenge in that the apparent gear mesh frequency is based on the gear's eccentric behavior and not on the gear mesh frequency alone. The observed gear mesh frequency is the gear tooth +1 vs. gear tooth, multiplied by the shaft rate.

Standard analysis techniques used on other gearboxes for shaft/gear, based on the time-synchronous average, were used. Forbearing fault detection, the envelope analysis was found to work well. For all components, a generalized health indicator was used to measure when maintenance actions were used.

Appendix

A.1 Example time synchronous average

```
Function  [tsadata, navgs,rpm] = tsaLinearInterp( data, zct, sr, ratio, ppr)
%[tsadata, navgs,rpm]=tsaLinearInterp(data,zct,sr,ratio,ppr,navgs)
%Inputs:
%     data:   time domain data in g's
%     zct:    zero cross time
%     sr:     sample rate
%     ratio:  gear ratio/pulse per revolution on the tach
%     ppr:    pulse per rev
%Output:
%     tsadata: time synchronous average data
%     navgs: the number of averages in the TSA
%     rpm: mean shaft rpm
%data = data - mean(data);
ndata = length(data);
dt = ndata/sr;     %sample length
rev = 0;
i = 1+ppr;
while zct(i) < dt && i < length(zct)-1
     rev = rev + 1;
     i = i + ppr;
end
% Define the number of averages to perform
navgs = floor(rev * ratio);
trev = zct(navgs*ppr) - zct(1);
rpm = navgs/trev*60*ratio;
% Determine radix 2 number where # of points in resampled TSA
% is at sample rate just greater than fsample
N=(2^(ceil(log2(60/rpm*sr))));
% now calculate times for each rev (1/ratio teeth pass by)
% resample vibe data using zero crossing times to interpolate the vibe
yy = zeros(1,N); %data to accumulate the resampled signal once per rev
ya = yy; %ya is the resample signal once per rev
iN = 1/N; %resample N points per rev
```

```
ir = 1/(ratio/ppr); %inverse ratio - how much to advance zct
tidx = 1; %start of zct index
  while (floor(zct(tidx)*sr) == 0)
      tidx = tidx + 1;
  end
    zct1 = zct(tidx);              %start zct time;
for k = 1:navgs
    tidx = tidx + ir;              %get the zct for the shaft
    stidx = floor(tidx)-1;         %start idx for interpolation
    dx = tidx - stidx;
    yo = zct(stidx);
    dy = zct(stidx+1)-yo;
    zcti = yo + dx*dy;             %interpolated ZCT
    dtrev = zcti - zct1;           %time of 1 rev
    dtic = dtrev*iN;               %time between each sample
    zct1c = zct1;
for j = 1:N
      cidx = floor(zct1c*sr);
      yo = data(cidx); y1 = data(cidx+1);
      x1 = zct1c*sr;
      xo = floor(x1);
      dx = x1-xo;
      dy = y1-yo;
      yaj = yo + dx*dy;            %simple linear interp
      ya(j) = yaj;
      zct1c = zct1 + j*dtic;  %increment to the next sample
end
    zct1 = zcti;
    yy = yy + ya;                  %accumulate the tsa per reve
end
tsadata = yy/navgs;                % compute the average
```

A.2 Example residual signal

```
  function [xres] = residualSignal(x, geartooth)
%[xres] = residualSignal(x, geartooth)
%Inputs:
%   x        :input TSA signal
%geartooth :array with number of teeth on a gear
%from Vercer
x = x(:)';
n = length(x);
n2 = n/2;
nHarmonics = 3;
X = rfft(x);                       %real fft - no conjugate
X(1) = 0;                          %DC is removed
X(2) = 0;                          % SO1 is removed
X(3) = 0;                          % SO2 is removed
nGears = length(geartooth);
```

```
for j = 1:nGears
    crtGear = geartooth(j);
 for i = 1:nHarmonics
      indx = 1+crtGear*i;
 if indx < n2                    %projection against running over the array
          X(indx) = 0;                    %gear tooth meash are removed
 end
 end
end
 xres = irfft(X);                % residual signal from the inverses real fft
```

A.3 Example of the narrowband, AM and FM analysis

```
 function [nb,am,fm] = narrowband(x, gt, BW)
%[nb,am,fm] = narrowband(x, gt, BW)
% x is the TSA
% gt is the number of gear teeth and
% BW is bandwidth, usually 25% of gt.
%Output:
%      nb: narrow band signal
%      am: amplitude modulated signal
%      fm: phase modulated signal
 X = rfft(x);
 lw = gt-BW; %calculate the band pass indexes
hi = gt+BW + 2;
 X(1:lw) = 0; %idealized filter
X(hi:end) = 0;
 nb = irfft(X);
 n = length(nb);
n2 = n/2;
X = fft(x); %take the Hilbert Transform
X(1:n2) = X(1:n2) * 2;
X(n2:end) = 0;
h = ifft(X); %Analytic Signal
 % Amplitude Modulation signal - am
am = abs(h);
 % Phase Modulation signal - fm
arg = unwrap(angle(h)); %take the argument
fm = arg - (arg(end)-arg(1))*linspace(0,1,n); %take the derivate
```

A.4 Example of the bearing envelope analysis

```
function [env,dty] = envelope(data,dt,lowf,highf)
% [env,dty] = envelope2(data,dt,nfilt,lowf,highf);
%Inputs:
% data      :data vector, time domain
% dt        :sampling time interval
% lowf      :low frequency limit of bandpass filter
```

```
% highf      :high frequency limit of bandpass filter
%Outputs:
% env :Envelope of data
% dty        : decimated sample rate
 n = length(data);
dfq = 1/dt/n;
idxLow = floor(lowf/dfq);
idxHi = ceil(highf/dfq);
D = fft(data);
idx = idxHi-idxLow + 1;
 D(1:idx) = D(idxLow:idxHi);
D(idx+1:end) = 0;
data = abs(ifft(D));
bw = highf - lowf;
r = fix(1/(bw*2*dt));
env = data(1:r:n);
 dty = dt*r;    %calculate the decimated sample rate
```

Author details

Eric Bechhoefer
GPMS International, Inc., Cornwall, VT, USA

*Address all correspondence to: eric@gpms-vt.com

References

[1] Cochran V, Bobak T. A Methodology for Identifying Defect Cycloidal Reduction Components Using Vibration Analysis and Techniques. Alexandria, Virginia: American Gear Manufacturers Association; 2008

[2] Stewart RM. Some useful data analysis techniques for gearbox diagnostics, Machine Health Monitoring Group, Institute of Sound and Vibration Research, University of Southampton, Report MHM/R/10/77 July

[3] McFadden PD. Detecting fatigue cracks in gears by amplitude and phase demodulation of the meshing vibration. Journal of Vibration, Acoustics, and Stress Reliability Design. 1986;**10**: 165-170

[4] Bechhoefer E, Butterworth B. A comprehensive analysis of the performance of gear fault detection algorithms. In: PHM Society Annual Conference. 2019

[5] Ma J. Energy operator, and other demodulation approaches to gear defect detection. Proceeding of the MFPT. 1995;**49**:127-140

[6] Kellar J, Grabill P. Vibration monitoring of a HU-60A main transmission planetary carrier fault. In: American Helicopter Society 59th Annual Forum. Phoenix, AZ; 2003

[7] Bechhoefer E, Van Hecke B, He D. Processing for improved spectral analysis. In: Annual Conference of the Prognostics and Health Management Society. 2013

[8] Hamrock B, Dowson D. Ball bearing mechanics. In: NASA Technical Memorandum 81691. 1981

[9] Rayleigh L. Theory of Sound. 2nd ed. London: Macmillian; 1894

[10] Timoshenko S. Theory of Plates and Shells. New York: McGraw Hill; 1940

Chapter 7

Probabilistic Risk Assessments for Static Equipment Integrity

Yury Sokolov

Abstract

The mechanical integrity of batch-produced machinery is successfully safeguarded using online condition monitoring and reliability theory principles. However, the integrity of nonreplaceable static equipment (pressure vessels, cranes, bridges, and other critical infrastructure) is still widely assured and managed using basic equations (e.g., safety factors and design loads), with no or little regard to the probabilistic nature of their operational damage. The gap between the deterministic "remnant life" assumptions and the probabilistic reality restrains the implementation of new asset integrity technologies (advanced condition monitoring and asset management) because these novel tools are not supported by a numeric cost/benefit analysis in many practical cases. The latter is impossible to implement confidently, while the probability of failure (PoF) versus time remains unquantified. The solution to this problem is holistic and logical: individual equipment integrity analysis now needs to be upgraded to the probabilistic terms at all the stages of life. Even well-known asset integrity technologies can help achieve this goal, providing that they are considered and utilized from the standpoint of harmonizing and aligning their outputs with risk owner's actual decision-making. This chapter shows real-life case studies to briefly illustrate how the existing integrity engineering tools can be advanced via further PoF considerations, in order to provide the outputs needed for a cost/benefit-based confident and compliant risk control.

Keywords: asset integrity, risk analysis, budget optimization, remnant life, probability of failure, corrosion, vibration, cracking, material fatigue, RBI, FFS, NDT, cost benefit

1. Introduction

A safe, reliable, and sustainable operation of an industrial plant is in the best interest of all the involved stakeholders. The sizes of modern hazardous process plants as well as their potential failure consequences can be enormous. One major challenge in their integrity risk management are the multiple equipment units experiencing specific operational and damage conditions, that is, one storage tank's corrosion damage is different from another due to different contents, one truck chassis cracking progress is different from another due to traveling on different roads, and one crane structure fatigue damage is different from another due to different histories of these cargo cycles. These examples explain the term "individual" equipment and render a batch reliability data or, especially, the "big data" not well applicable to them due to unit-specific load and damage spectra acting in a real operation.

Historically, the first approach to safeguarding equipment integrity was reactive: failures were rectified as they happen, but it was not a responsible strategy for hazardous equipment. A transition to proactive maintenance occurred over the automotive industry development, as we are familiar from the time/mileage-based car servicing. That solution obviously improved the reliability, but its cost control efficiency in practice can vary. In parallel, statistical quality control principles were implemented in manufacture to ensure a uniform endurance of production batches and facilitate the reliability theory [1] applications.

In contrast, there was not such a scientific breakthrough in the domain of static equipment, which is hardly maintainable or replaceable, nonredundant, and not suitable for collecting failure statistics due to high consequences thereof. The static equipment integrity is traditionally addressed via time-based (fixed interval) diagnostics, often using visual in-service inspections, as in the oil and gas industry. In this way, an inspector takes responsibility for the equipment fail-safe operation during a future fixed term, while no in-depth analysis is actually done for a scope damage potential (mostly a form of corrosion and cracking or, more occasionally, metallurgical changes and material properties degradation).

The potential of missing or misinterpreting a damage condition was effectively alleviated by adopting the risk-based inspection (RBI) principles two decades ago. The main idea of RBI is proportioning the risk control efforts to the individual risk levels, that is, prioritizing the equipment units for reinspections according to their relative risks across the plant. But how to measure risk levels without excessive analysis budgets in a context of a large plant? The widely adopted robust solution is the semiquantitative (Semi-Q) RBI, which uses corporate risk matrices to unify and compare relative failure risks unit by unit:

$$\text{Risk} = \text{LoF} \cdot \text{CoF} \tag{1}$$

where LoF is the likelihood of failure and the CoF is the consequence of failure.

The size of the risk matrices is usually 5×5, and the LoF and CoF enter Eq. (1) as dimensionless multipliers ranging from 1 to 5; thus, the product risk varies from 1 to 25. CoF ratings are mapped from considering safety, financial, and environmental impacts of the unit failure, which are confidently assigned using plant operations' personnel knowledge. LoF ratings are mapped from the anticipated "remnant life" (RL). In corrosion problems, RL is calculated from dividing a corrosion allowance CA by a corrosion rate CR:

$$\text{RL}\left[\text{years}\right] = \frac{\text{CA}\left[\text{mm}\right]}{\text{CR}\left[\frac{\text{mm}}{\text{year}}\right]} \tag{2}$$

It is paramount that the risk ratings from Eq. (1) are dimensionless, and their evolution in the future remains unknown. This simplification disables a numeric cost/benefit analysis in terms of dollars and fatalities, and, thus, the asset management aspirations. In turn, it provides no justification for implementing advanced nondestructive testing (NDT) tools, as the figures entering Eq. (2) are available from basic and low-cost ultrasonic thickness (UT) gauge inspections. A numeric comparison of risk control options is not supported either.

Other fitness-for-service (FFS) problems [2], such as fatigue life, crack propagation intervals, tolerance to mechanical defects and imperfections in a wide spectrum of stress, and environmental conditions, all involve some form of stress field measurement or modeling. Stress modeling can be done using finite-element analysis (FEA), with an added benefit of reducing an uncertainty in stress concentration factors (SCF) and of performing a relatively quick analysis even for very complex geometries. But again, FFS and FEA studies often output constant figure "remnant lives"; thus, the above limitations apply.

As a matter of big picture, there are many advanced integrity assessment technologies developed to date, but they are not well aligned to each other or to the common umbrella of the asset management concept [3], by the major reason of providing single-figure outputs. Namely, a single-figure "remnant life" does not exist. What exists in reality is an individual probability of failure (PoF), which grows over time due to the mechanical damage accumulation. This applies to corrosion, fatigue, and other mechanical strength problems. Next examples show how a simple transition from the single figure to the PoF(t) function contributes to the risk owner's decision-making process both numerically and qualitatively, thereby aligning the asset integrity technologies together to provide numeric cost/benefit outputs.

2. PoF estimates in harmonic vibration

Over the past century, machinery has become much more powerful and high speed. More power leads to more energy losses, which are dissipated mostly in the forms of heat, vibration, and noise. Mechanical excitation from reciprocating machinery is not the only vibration source in a modern plant. Acoustically induced vibration (AIV) and flow-induced vibration (FIV) also occur in power circuits of compressors and pumps. An excellent overview of these vibration mechanisms is given in the UK Energy Institute Guidelines [4]. FIV and AIV often occur at no flow piping branches, such as small bore fittings (SBF) (**Figure 1**), designed for process probes, ancillary access, or for draining and venting purposes.

Real-life case: High vibration levels were measured on SBFs of 11 compressor pulsation bottles during a gas plant commissioning. AIV velocities of up to

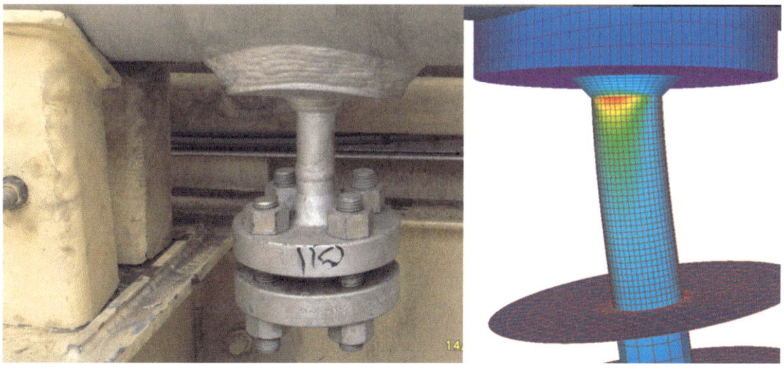

Figure 1.
A small bore fitting (SBF) and its FEA model.

Unit	A	B	C	D	E	F	G	H	I	J	K
TTF [hours]	341	382	188	385	373	505	449	290	50	299	455

Table 1.
SBF failure statistics; TTF stands for time to failure.

29.5 mm/sec root mean square (RMS) at 150 Hz were recorded using portable vibration equipment. These figures were screened using the chart of [4] and, accordingly, classified as a "concern" region. The commissioning was continued, and all 11 pulsation bottles failed within 500 hours (**Table 1**).

In this example, the SBF tends vibrating at its natural frequency about the zero mean (M) level harmonically, and its displacement peaks follow the Gaussian probability distribution. The RMS vibration displacement (of 31 micron here) is equal to one standard deviation (SD) of this random displacement. This displacement can be converted into the weld root bending stress amplitude (see the red spot in **Figure 1**) even manually—using simple beam theory of materials strength in view of this particular geometry simplicity. The nominal stress amplitude of 12.2 MPa RMS was estimated, and the whole stress spectrum was reconstructed analytically to obey a zero-mean Gaussian law having this very SD value.

The nominal bending stress formulation is compatible with the BS 7608 [5] standard material fatigue data (category F), which data were formerly obtained from large-scale testing or real weld details. Other standards (ASME VIII [6] and EN 13445 [7]) require more complex stress formulations, which would normally involve finite-element analysis.

In the risk owner's context, the problem is: "How long will it last?" Answers can vary:

1. Using constant stress amplitude (such as 1·SD, 2·SD or 3·SD) with single-figure standard fatigue data is here typical, but an incorrect approach. Material fatigue analysis does not tolerate simplifications and/or factors due to the high nonlinearity of the fatigue life in function of the stress level. If a structure is subjected to a spectrum of stresses, then each tower of the stress histogram has to be input into the fatigue analysis, and the total damage should be calculated as a sum of contributions from each tower according to the Miner's rule [see Eq. (4)].

2. Using the whole stress spectrum (as suggested just above) is a step forward indeed, but in conjunction with a single-figure fatigue strength value, it will lead us to the same pitfall: a single-figure remnant life output with an unknown risk evolution in time. The solution is found in the fatigue damage physics: Material strength is a random variable statistically independent from the live stress spectrum it experiences, as illustrated by the two probability density functions (PDFs) in **Figure 2**. This simple schematic of the load and resistance interaction can be found in reliability theory textbooks (such as [1]) and is often called "bell shape" curves.

3. Since these two variables, *P(stress)* and *P(strength)*, are statistically independent, a simultaneous occurrence of a certain stress level x and a certain strength level x is a product of their PDFs. The "Monte Carlo" method [8] can be used for generating such random variables. An analytical expression for determining the PoF is, similarly, a product of the two probabilities, but the cumulative density functions (CDFs) are applicable instead:

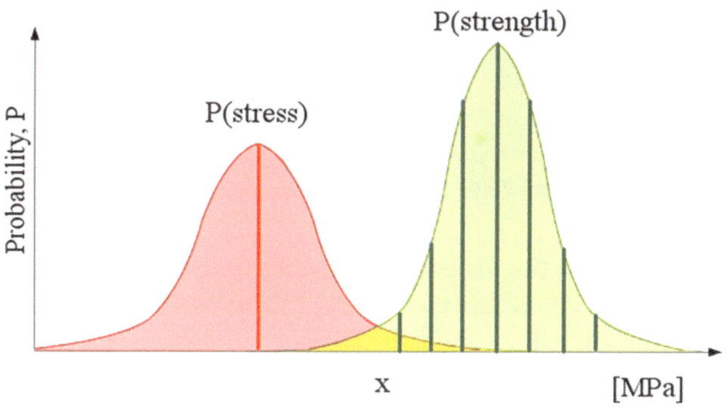

Figure 2.
"Bell shape" curves showing the product of probabilities.

$$\text{PoF} = P\,(stress > x) \cdot P(strength < x) = P(stress > strength) \quad (3)$$

In this example, the reconstruction of the Gaussian stress spectrum enabled the use of the whole red "camelback" shape from **Figure 2**. The spread of fatigue strength properties is naturally available from specimens testing data and manifests itself as a change in a fatigue curve position as the number of standard deviations (SDs) around mean (M) is varied. Thus, replacing the green shape in **Figure 2** by a histogram of discrete *P(strength)* levels and repeating the fatigue calculations over the whole stress spectrum provides a robust solution for approximating the PoF(t).

The above solution for the analysis upgrade is not only reflecting the damage physics more precisely (than a "single-figure" route), but also enables seamless cost/benefit considerations made from converting the PoF(t) (left in **Figure 3**) into \$risk(t) and safety_exposure(t). The PoF(t) function multiplied by a likely financial impact of the failure gives the cost of risk in dollar terms (left in **Figure 2**). The likely \$100,000 cost of failure due to delayed commissioning was applied here. The clearly visualized growth of the dollar risk versus hours in operation suggests that the risk should have been mitigated within few days. Yet, another effect of this failure can be workers' safety exposure, to be safeguarded by the owner via setting a PoF limit, example of which is shown in Section 5.2 (right in **Figure 11**)

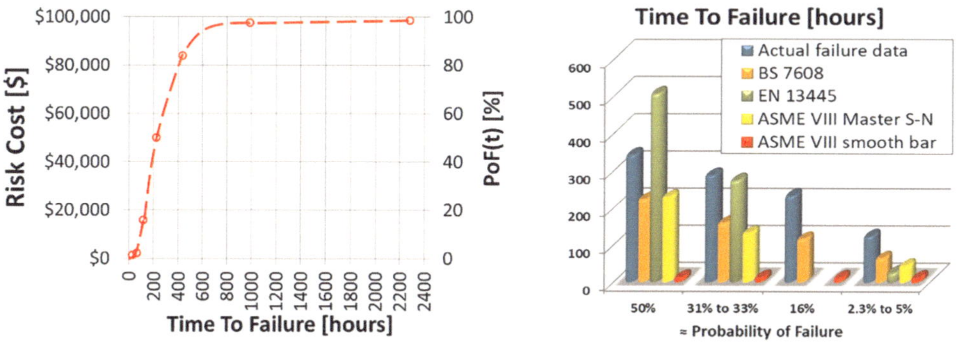

Figure 3.
SBF-estimated PoF(t) *function and a benchmark of popular fatigue standards.*

BS 7608 *P(strength)*	1.4%	2.3%	16%	50%	84%	97.7%	98.6%
Predicted *TTF* [hours]	29	61	112	219	431	982	2276

Table 2.
Example of PoF(t) predictions for a harmonic vibration case.

According to **Figure 2**, the stress histogram was used with fatigue curves at seven (M ± i·SD) levels of the weld detail fatigue strength results with the output shown in **Table 2**.

Some final remarks to this study can also be useful for other practical applications:

4. Particulars of fatigue methodologies vary across the standards, as shown on the right in **Figure 3**. A benchmarking study has been done for this problem and published on the ResearchGate network [9]. It has concluded that the BS 7608 [5] standard in conjunction with its simple input data requirements performed best in this particular problem, showing slightly conservative outputs. Notably, if two standards output different figures, then one would be closer to the reality and another further away from it. The benchmark in **Figure 3** quantifies this example effect. The reasons for fatigue methodology differences across similar application domain standards were earlier investigated in yet another ResearchGate paper [10].

5. The mean time to failure (TTF) in this example is 338 hours at 150-Hz frequency, that is, 1.8×10^8 stress cycles, or a "gigacycle fatigue" (GCF) regime. The term "gigacycle" was introduced by the fundamental research published in [11, 12]. Its major conclusion was that a "*fatigue limit beyond which fatigue failures of steels do not occur*" does not exist as a physical phenomenon. Fatigue failures of steels do occur beyond 10^8, 10^9, and 10^{10} load cycles even at small stress amplitudes. Failure data **Table 1** also confirm this. Modern standards extrapolate fatigue testing data to 10^8–10^9 cycles, and this approximation showed itself well applicable to vibration.

3. PoF predictions from strain gauging data

3.1 Constant amplitude response

Strain gauges [13] (left in **Figure 4**) can be attached to structures to record mechanical strains and further convert them into material stresses. This technique provides the most reliable information on the live stress spectra in real operation of industrial equipment. Care should be taken to ensure that the recorded process is representative of the dominant operation.

This real-life example deals with temperature- and pressure-induced stresses in a glycol pump pulsation dampener nozzle. The pump run-up cycle stresses were strain gauged in a typical pump "mission," as shown in **Figure 5**.

Accordingly, the bending stress range of up to 56 MPa occurs in each run-up/shut-down cycle due to the increase in pressure by 112 barg and the piping heating up from 27°C to 70°C, which is representative for this particular plant process. There is one major stress cycle of this magnitude occurring during each run-up event; thus, the stress spectrum (**Figure 2**) collapses into a single vertical red line in lieu of the whole red bell shape *P(stress)*.

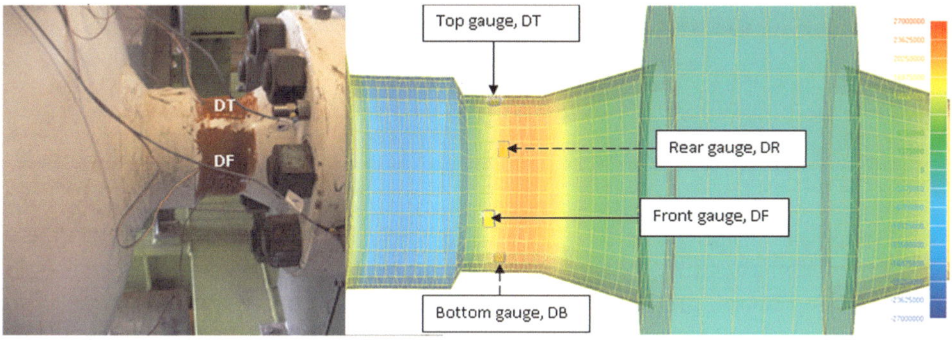

Figure 4.
Strain gauges attached to a pressure vessel nozzle and its FEA model.

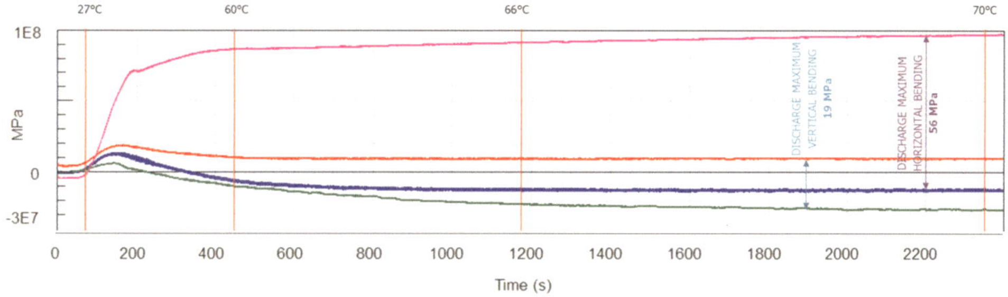

Figure 5.
Nozzle stresses recorded during the glycol pump run-up cycle.

Statistical variation of the material (SA 106 B) properties still needs to be considered. This is done similarly to the previous example via usage of fatigue curves corresponding to varying probability levels of the material fatigue strength (green vertical lines in **Figure 2**).

One nuance here is that strain gauges cannot be positioned exactly on stress "hot spots" as the latter usually occur at structural discontinuities visible in **Figure 4**. The pressure vessel design code EN 13445 [7] contains a provision for stress extrapolation in such cases using readings from two locations of strain gauges (or of an FEA mesh). The above measurement had only one strain gauge at each location; however, an FEA model of the dampeners (right in **Figure 4**) provided the figures of stress gradient along the nozzle length helpful for such an extrapolation. It is evident from **Figure 4** that the stress concentration effect in this case does not exceed 1.25, and thus, the extrapolated stress range should not be more that 70 MPa (zero to peak). The weld detail classifies as the Category 32 (fillet and partial penetration welds) fatigue curve given in [7]. By varying the number of standard deviations (SDs) of the CAT 32 fatigue data, we get the varying number of cycles to failure straight away.

Since the frequency of the pump run-up/shutdown cycles is no more often than once a day, the number of cycles in **Table 3** maps directly into the number of days, that is, 288 years at the lower bound failure probability. Hence, the equipment should not fail by the nozzle fatigue mechanism until the end of the offshore platform life, providing that the recorded constant amplitude conditions were representative for the whole operation of the pump.

EN 13445 PoF	**0.0135 (M – 3·SD)**	**0.023 (M – 2·SD)**	**0.156 (M – 1·SD)**	**0.50 (M – 0·SD)**
Cycles to failure	1.07e5	1.24e5	1.5e5	1.2e6

Table 3.
PoF(t) prediction in the nozzle strain gauging case study.

This example simplicity is due to the actual constant amplitude loading. It shows how the probabilistic integrity analysis unambiguously supports the asset management decision-making process. One remaining safeguard is performing a penetrant inspection (PI) of the nozzle to ensure that there are no cracks from other reasons (transportation, impacts, etc.).

3.2 Variable amplitude response

This example illustrates a more complex situation where strain gauging provided a true stress spectrum for a mining truck tray hot spot. A total of 18 potential hot spots were strain gauged using triaxial rosettes during a typical truck mission involving: loading rocks in the tray, travel, emptying, and returning to mine site several times during a 7-hour-long shift. The most critical location of the tray was identified as a result and is shown in **Figure 6**.

Signal processing software was used for the analysis, and the output fatigue damage spectrum is shown in the left of **Figure 7**. The maximum principal stress range was used, as the fatigue crack growth is governed by the maximum stress component opening the crack.

The majority of fatigue damage in the left of **Figure 7** occurred in the low-stress area; however, few spikes up to 290 MPa were recorded infrequently during the tray loading. The whole damage spectrum is a good illustration of a variable amplitude fatigue loading, and the damage introduced by each stress range *i* is calculated according to Miner's rule [14]:

$$D = \sum_{i=1}^{n} \frac{n_i}{N_i} \tag{4}$$

where n_i is the number of cycles brought at the ith stress level and N_i is the number of cycles to failure at this very stress level obtained from a relevant fatigue curve.

Figure 6.
Mining truck tray and its critical location identified from strain gauging.

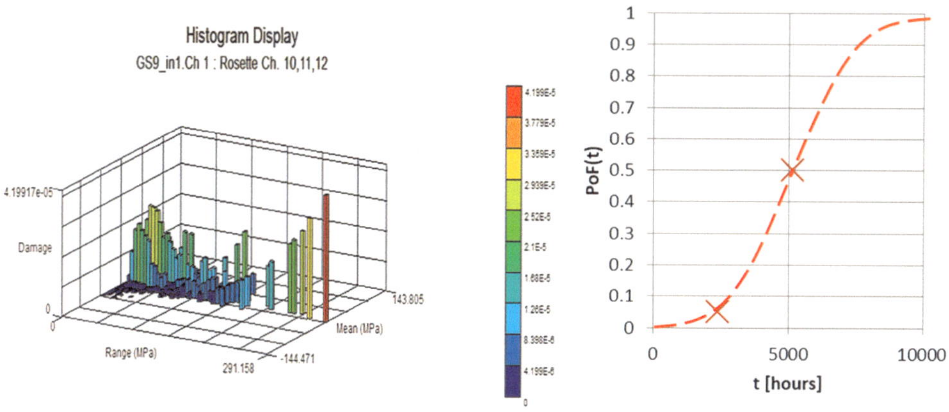

Figure 7.
Tray damage spectrum accumulated during one shift and the PoF(t) estimate.

BS 7608 PoF	0.05 (M – 2·SD)	0.50 (M – 0·SD)
Time to failure [hours]	2374	5150

Table 4.
PoF(t) prediction for the mining truck tray hot spot.

Unlike the previous example where the stress field extrapolation was required by the standard [7], the present example used BS 7608 fatigue data [5]. The philosophy of the latter is slightly different: real weld details were tested for fatigue with the output of nominal structural stresses. In turn, nominal stresses are used with the fatigue curve of [5], e.g., those stresses reasonably away from hot spots, as it was attempted to collect by placing rosettes at a small distance from the stress raisers (refer **Figure 6**). The BS 7608 detail Category G Class 5.5 fatigue data were used at two levels of its probability (**Table 4**).

Material testing data for the M and M – 2·SD levels can be found in technical literature most often, and these two points can be used to approximate the PoF(t) S-shaped curve up to the 50% level even by a smooth curve manual fitting, considering that the third point is

$$\mathrm{PoF}(t = 0) = 0 \tag{5}$$

The vendor's guarantee on the tray life was 20,000 hours, and this worst-case location was recommended for reinforcement as an outcome from the above analysis. A self-explanatory picture of the PoF(t) function was obtained from a manual fitting of a typical S-shaped cumulative density function (CDF) to these two estimates, as shown in the right of **Figure 7**. Using more PoF levels would further improve the PoF (t) curve shape accuracy if needed. This is an example of a design support made from the records of a pilot exemplar operation.

4. PoF considerations in fitness-for-service problems

The term fitness for service (FFS) is used where damage in excess of a design tolerance has already been found in the equipment, and this analysis aims at replying two questions:

a. How critical is the defect at the moment of its characterization?

b. How long will that equipment last in view of this defect future growth?

The first question triggers a pass/fail or a screening-type output, and the second drives a fixed "remnant life" figure in many studies. While the FFS methods do use empirical methods (such as crack growth laws), applications of FFS analysis are unfortunately narrow. This is mostly due to their complexity and timing, while risk owners need prompt decisions in such critical situations. The same upgrade idea can be used to output the damaged equipment PoF versus time and add more value through visualizing the risk evolution.

4.1 Crack propagation problems

Port cranes (left in **Figure 8**) showed three failures by fracture of the boom top shelf (right in **Figure 8**), which resulted in catastrophic consequences. Since then, the manufacturer has reinforced the boom design. However, a life extension decision was required in the late 1990s, and that decision needed a scientific substantiation in view of potential failure implications.

As it was mentioned in the introduction, cranes are highly individual structures in the sense of their loading, and a screening using a conventional fatigue theory showed that a "generic" port crane has a life expectancy of 25 years ± 30 years spread, which outcome is not practical.

The solution was in adopting the damage tolerance approach: cracking inspections to be implemented at individual intervals. If cracks are not found, then it is assumed that a crack of a nondetectable length (less than 5 mm) is nevertheless present. A life extension is then warranted for a safety factored period needed by that crack to grow to a critical size. This scenario required only basic visual inspections, but had a good potential to control the risk. An earlier application of a similar method for bridges life extension can be found in [15].

Figure 8.
Port crane structure and an FEA model of its cargo boom.

The relevant science apparatus is the fracture mechanics empirical laws of crack growth detailed for example in the BS 7910 FFS standard [2]. Since this theory is a rather uncommon specialist knowledge, a simplistic introduction follows here.

In function of the material, temperature, and the strain rate, there is a variable-size plastic zone at a crack tip. Thus, the stresses there are singular, and the fatigue theory term "stress range" is not straight applicable to predict the crack growth rate. Instead, a stress intensity factor (SIF) range ΔK [MPa$\cdot\sqrt{\text{m}}$] is used to correlate a "nominal" stress range $\Delta\sigma$ away from the crack tip with the empirical crack behavior:

$$\Delta K = F \cdot \Delta\sigma \cdot \sqrt{\pi \cdot a} \tag{6}$$

where F is a geometry constraint correction and a is the half-length of the crack.

Cracks grow nonlinearly; they accelerate as they grow starting from microns per cycle and ending with a catastrophic growth rate. The empirical Paris law approximates this process:

$$\frac{da}{dN} = C \cdot \Delta K^m \tag{7}$$

where the left-side derivative is the crack growth rate, N is the number of cycles, and C and m are material properties–probabilistic variables known from statistical treatment of test data.

Using mathematical transformations, the system of Eqs. (6) and (7) yields the crack length increase (from size a_i to a_{i+1}), which can be estimated in each stress cycle, one after another:

$$a_{i+1} = \sqrt[1-\frac{m}{2}]{C \cdot \Delta\sigma^m \cdot \pi^{m/2} \cdot F^m \cdot \left(1 - \frac{m}{2}\right) + a_i{}^{1-m/2}} \tag{8}$$

Eq. (8) is suitable for simulating the crack growth cycle by cycle using the Monte Carlo method. Nuances are numerous, but two of them are sometimes overlooked in practice:

- Cracking often initiates in heat-affected zones (HAZ) of welds, where residual tensile stresses originate from welding and do affect the crack tip opening.
- Structural stress gradients affect the nominal stress range $\Delta\sigma$ as the crack grows.

To include these stress gradients, a cycle-by-cycle Monte Carlo simulation has been performed, and the results compared with the output of the simplified equations below, which estimate the total (e.g., integral) number of stress cycles N_C necessary for the crack to grow from an initial size a_0 to the critical size a_C:

$$N_c = \frac{1}{C \cdot \Delta\sigma_{eq}^m \cdot \pi^{m/2} \cdot F^m \cdot \left(1 - \frac{m}{2}\right)} \cdot \left[a_c^{1-\frac{m}{2}} - a_0^{1-\frac{m}{2}}\right] \tag{9}$$

where F is the geometry constraint correction, C and m are the probabilistic fracture resistance characteristics of the material (we will vary them just below), and $\Delta\sigma_{eq}$ is the equivalent nominal stress range derived from the measured stress spectrum as follows:

$$\Delta\sigma_{eq} = \sqrt[m]{\sum_{i=1}^{j}\left(\Delta\sigma_i^m \cdot f_i\right)} \tag{10}$$

where $\Delta\sigma_i$ is an ith tower of the stress spectrum histogram and f_i is its occurrence frequency.

The Monte Carlo validation proved Eqs. (9) and (10) being correct and underestimated the crack propagation life by some 30% compared to the stress gradients included. The equivalent nominal stress range $\Delta\sigma_{eq}$ = 99 [MPa] resulted from strain gauging and FEA for the original, not reinforced design of the boom. The left plot in **Figure 9** reads as the number of daylong crane shifts in function of a detected crack length in various crane missions (e.g., cargo cycles). Consider a 5-mm-long crack at the hot spot of concern: the number of shifts till failure varies from 22 to 123 depending on the duty cycle severity.

Now, let us enrich this research project from the early 2000s by considering two probability levels of the steel fracture resistance parameters C and m, similarly to the previous example (**Table 5**).

The account of material properties variation also gives an order of magnitude change in life predictions, resulting in 112 shifts using the mean properties, as opposed to 22 shifts resulted from the lower bound data (taken for the worst-case cargo cycle—the brown curve in the left of **Figure 9**). Similarly, manual fitting of an S-shaped curve to these two data points produces a smooth PoF(t) function (right in **Figure 9**) to visualize the failure chance.

Multiplying the PoF(t) by the likely cost of the crane replacement and the penalties involved will estimate the $Risk($t$) for a cost/benefit decision-making. Safety implications here are also severe and can likely lead to one or two fatalities (one docker and one crane operator). Providing that the risk owner has a safety limit, it should be used as a cutoff on the PoF.

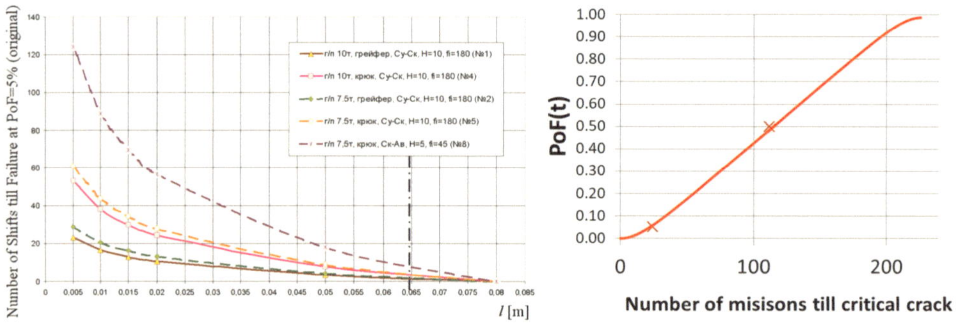

Figure 9.
Crane boom PoF(t) due to crack growth from 5 to 80 mm length in a worst mission.

P(fracture properties)	5% (original study)	50% (present study)
C	5.97e–11	1.44e–14
m	2.25	4.72

Table 5.
Carbon Steel (St38b2) fracture resistance parameters at two levels of their probability.

4.2 Other damage mechanics relevant to the FFS scope

Getting back to the FFS scope of problems [2], in majority of cases, these are:

1. Fatigue and crack propagation governed damage (addressed above)

2. Creep (empirical analysis apparatus generally similar to the present fracture mechanics example, which is suitable for a similar probabilistic analysis approach)

3. Corrosion and/or erosion driven material wastage (discussed in next section)

4. Gross defects affecting the distribution of loads and strains (modeled by FEA and then analyzed versus operational stress spectra similarly to examples in Section 3).

Thus, the majority of operational damage cases can be quantified using the PoF(t) strategy.

5. PoF estimates from corrosion data

The problem of corrosion failures, surprisingly, is the most technically challenging for estimating the PoF(t) function. This is because spatial distributions of corrosion damage are also probabilistic, further aggravated by the practical inability to inspect 100% of the equipment surface. The challenge of equipment internal corrosion risk control is major in petrochemical industries, and failure implications are severe, as well as the inspection costs.

5.1 The "state of the art" in assessing pressure equipment corrosion

The most natural and straightforward corrosion risk analysis methodology was outlined in the introduction to this chapter and is called "Semi-Q" risk-based inspection (RBI) planning. It is very robust for large plants and does output a relative risk ranking. However, the dimensionless risk levels are not aligned with a numeric cost/benefit analysis and personnel safety demonstration in this context and, thus, require an upgrade.

Another popular RBI methodology API RP 581 (refer [16] for technical background) is used in most RBI software. For a simplistic explanation, their POF values originate chiefly from

$$\text{PoF} = GFF \cdot DF = GFF \cdot \left(\frac{1}{EFF}\right) \cdot f\left(\frac{AGE \cdot CR}{THK}\right) \tag{11}$$

where *GFF* is a constant generic failure frequency by equipment and damage morphology, *EFF* is an inspection efficiency factor reduced for each next year by 10%, and f is tabulated as a function of the parameters in brackets: equipment age (*AGE*) at the inspection time, the estimated corrosion rate (*CR*), and the wall thickness (*THK*) available for wastage.

The meaning of f is, thus, a ratio of the wall loss (*WL*) "as inspected" to the remnant *THK*:

$$AGE\ [\text{years}] \cdot CR \left[\frac{mm}{\text{year}}\right] = AGE\ [\text{years}] \cdot \frac{WL\ [\text{mm}]}{AGE\ [\text{years}]} = WL\ [\text{mm}] \tag{12}$$

The recent API 581 editions change from second to third refined the *THK* calculations to consider the minimum required wall thickness (MRWT) parameter and that increased the conservatism (refer to the left chart in **Figure 11**). API 581 offers useful data for non-age-related damage mechanisms, but its thinning assessment method has two strategic pitfalls:

1. using generic constant frequencies *GFF* for individually damaged equipment

2. using a single "worst-case" corrosion location, thus neglecting the rest of them.

The latter is a clear indication of distorting an actual PoF because a pool of thickness readings did contain the intrinsic corrosion distribution information. This information cannot be restored if it was collapsed into a "worst-case" data point; hence, an analysis done from a single location will not produce a true PoF, as one of probabilistic distributions was ignored.

Quite apart stands the DnV-RP-G101 [17] RBI methodology, which extensively uses PoF terms for age-related (time-driven) and non-age-related (process-parameter-driven) damage mechanisms. The terms are linked to the quantitative consequence assessment, and three levels of assessment detail are recognized too. One major simplification, again, is using generic PoF varied by a damage mechanism type there. PoF data in [17], thus, enables PoF estimates with no inspection data involvement whatsoever. This is useful for design, but quite confusing for assessment purposes. We observe the same attempt of generalizing failure probabilities for individual equipment and neglecting the true spatial distribution the damage. Hence, same as above pitfalls 1) and 2) apply in the DnV-RP-G101 method too.

Perhaps, the most comprehensive statistical treatment of corrosion data is outlined in Appendix B of the Nonintrusive Inspection guideline DnV-RP-G103 [18]. This guideline resulted from the HOIS Joint Industry Project to assist implementation of advanced NDT tools (such as large coverage corrosion mapping) in the oil and gas industry. It introduces the extreme value analysis (EVA) [19] applications to large samples of corrosion data. In brief, the data points x are first statistically plotted on a probability paper having custom scaled axes; second, a probabilistic distribution CDF (x) is fitted and is then extrapolated to a "survivor function" SUR(x) using the ratio of the total equipment area to the inspected area [20]:

$$\text{SUR}(x) = [1 - \text{CDF}(x)]^{\text{Total Area/Inspected Area}} \tag{13}$$

Finally, a "worst-case" reading is found from the survivor function at a target level of its occurrence probability, say 1%. Thus, the whole data are collapsed into a single point again.

Seemingly, there is psychological antagonism in such a scenario: advanced NDT providers aim supplying more and better data, but collapse it to a single value, as they are asked by the risk owner to produce a "worst location." This is because RBI methods require a single location for a corrosion assessment, and thus, advanced NDT applications add little more value.

5.2 Proposed method for corrosion risk analysis

The solution proposed here (and previously reported at few industrial conferences) is using the same bell shape curves product principle (right in **Figure 10**) for corrosion risk assessments. In contrast to the above methods, it retains all the relevant inspection data points and uses the corrosion damage distribution “as is” (left in **Figure 10**), without any fixed value extrapolation or user factoring involved:

The brown points are the corrosion data “as measured” with a Gumbel distribution fitted (dashed line), and the green curve is the cumulative density function (CDF) of this individual corrosion distribution. The probability of failure in this case is also a product of two events:

$$PoF = P(THK\ occurence) \cdot P(Failure\ at\ that\ THK) \tag{14}$$

The probability of failure at a certain thickness level is also equipment individual. It can be quantified as in the above examples or even more simplistically. The PoF in Eq. (14) is instantaneous at the moment of inspection. To assess the PoF(t) evolution in time, the evidential corrosion rate is simulated for the future time instances, and that effectively shifts the green bell shape in **Figure 10** to the left. The blue overlap area grows, and so does the PoF obtained from Eq. (14).

A PoF(t) function predicted from a real-life pressure piping case study is shown on the left of **Figure 11** (solid blue line). The safety exposure limit of one fatality in 1000 workers per year is shown by the red-dotted line. Their intersection means the safety

Figure 10.
Product of probabilities in corrosion problems.

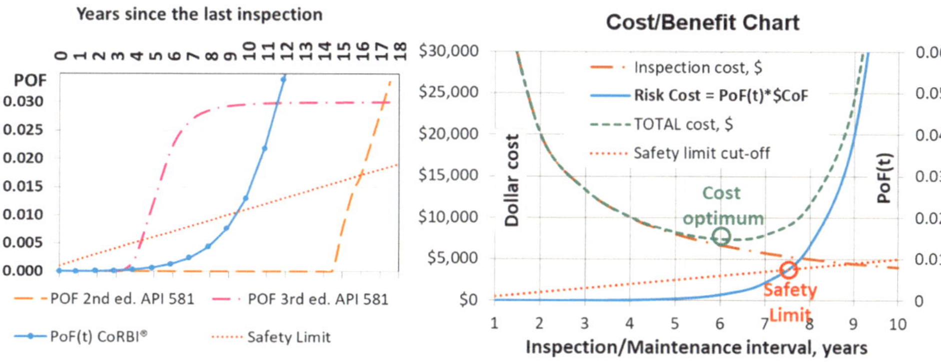

Figure 11.
Corrosion PoF(t) predictions benchmark and cost/benefit analysis.

limit breach. Operation past this time instance will not be compliant with it. Finally, the transition from a PoF(t) to the risk dollar cost is multiplying PoF by the anticipated total cost of the failure consequences (near $1 million here due to nonredundancy and collateral damage potential). This is shown by the solid blue line in the right of **Figure 11**, with the cost being read from the left vertical axis, and it obviously increases over time.

A surprisingly common confusion is that inspections affect PoF or risks. This is not the case until actual risk controls have been implemented following the inspection and do physically minimize or mitigate risks, similarly to the resource restoration in the reliability theory [1].

The dashed-dotted line depicts the cost of all inspections done, totaled toward the end of equipment life, in function of the variable inspection interval (horizontal axis). The sum of the solid and dashed-dotted lines is the total cost (of risk and inspections), which has a minimum at 6 years since the last inspection here. It should be used to reinspect or set other relevant risk controls (replacement, barriers, and process changes), providing that they occur prior to the safety limit breach at 7.5 years in this example. Otherwise, the safety limit must prevail.

The cost/benefit plotting shown on the right of **Figure 11** is especially useful for building effective asset management frameworks, as it facilitates an unambiguous budget allocation made from the numeric figures of risk exposure and their comparison with mitigation costs.

6. The upgrade potential and way forward

The above material illustrates an integrity analysis upgrade potential resulting from the new strategic premise that every operational integrity assessment should output PoF(t) to assist justified decision-making regarding individual equipment maintenance and risk control.

The asset management concept [3] offers a common umbrella for all integrity risk control decision-making, including the adoption of advanced condition monitoring (CM) tools and digitalization technologies on the basis of their cost and safety control efficiency. In turn, the latter is assisted by providing an adequate level of data analysis using the PoF(t) strategy, while this very strategy also enables the cost/benefit charting. In this way, the presented research and development was not occasional or voluntary, but triggered by the challenges in implementing advanced technologies (RBI, FFS, FEA, and NDT). Therefore, this chapter aimed at showing the big picture of these problems and our holistic PoF(t) solution to them.

The methodology is regarded complete as the following has been achieved to date:

- The concept of estimating PoF(t) as the product of two statistically independent events was applied to a range of damage physics, as illustrated above.

- The shown real-life examples of all the output predictions were consistent with operational experience and were well agreed upon by experienced professionals in this field, e.g., inspection and integrity engineers responsible for those particular problems troubleshooting. No artificial factors were used, but these studies have output very sensible figures. This reinforces the validity of the methodology.

- The transition to the cost of risk and safety exposure tolerance was made using likely consequences of failure. Estimating CoF is usually done at ease by the relevant site personnel. A further refinement of CoF is feasible using a Layers of Protection Analysis (LOPA) if this is warranted by risk levels and control systems.

- The rightful concept of risk-based integrity control was applied to all the studied problems. In other words, the level of analysis should be proportional to the problem criticality. The PoF(t) concept is relevant to high criticality problems and interacts synergistically with simpler practices relevant to lower risk objects. In this way, the analysis depth can be escalated through several levels as the risk estimate is being refined and does indicate a requirement for an escalation.

- The methodology also does not contradict with any modern inspection and risk analysis standards, but supplements their capabilities via more advanced data analysis and aligns the particular data collection and analysis apparatus with the asset management aspirations of cost and risk control.

- The implementation of the method does not demand for an instant step change in condition monitoring tools, as wide spread technologies (spot check UT, strain gauging, and vibration accelerometers [13]) are sufficient to support its initial implementation as shown above. In turn, this implementation will provide a numeric cost/benefit basis for advanced CM tool implementation consideration.

- The PoF(t) concept is based on the actual damage physics, and since a particular material behavior (material fatigue, crack propagation, and corrosion mechanisms) describe the nature laws, their application is universal across industries and life stages. This is a holistic solution able to support asset integrity in any industry.

- Finally, the upgrade is not too cumbersome technically, as the most labor in static equipment operational integrity assessments is spent on measuring and modeling the damage phenomena, while the addition of multi-PoF analysis only requires repeating certain calculations few times and visualizing the new results.

And the way forward is obviously to expand trials of this methodology across industries, work through particular nuances where required, and validate its application benefits. The concept implementation now became feasible thanks to the cross-industry adoption of precise measurement techniques applicable to integrity problems, although not yet fully realized.

One misconception found in practice is applying design premises to operational integrity assessments. The "design life" concept has another purpose, and it is still open for further improvements [15] via evidential data. Reliable data originate from *in situ* measurements ever expanding in their capabilities over the past two decades. The only major challenge in implementing more and better monitoring is the financial justification, which can be resolved using the above methodology to maintain the static "nonmaintainable" equipment.

To conclude, the following quote from Galileo Galilei outlines the general research concept eventually reinforced here: "*Measure what is measurable, and make measurable what is not so*."

Acknowledgements

The author is sincerely grateful to his teachers who guided his work on the thesis (section 4.1). He also very much appreciates the hard work of field engineers, who were collecting the live data (Sections 2 and 3) during his times at SVT Engineering Consultants (Perth). The R&D work on implementing the PoF(t) and Risk Cost terms into integrity assessments was undertaken by Quanty Pty. Ltd. at own expense with no finance or influencing by others.

Conflict of interest

None exist.

Disclaimer

The information in this chapter aims at highlighting a big picture of the probabilistic analysis process and its implementation potential made in a simple language. It does not show all the nuances or technical details of these examples. Since the scope problems are individual, the above data and simplified equations should not be applied to other individual equipment cases. We disclaim any liability resulting from an application of this information by others.

Author details

Yury Sokolov
Quanty Pty Ltd, Brisbane, Australia

*Address all correspondence to: contact@quanty.com.au

References

[1] Carter ADS. Mechanical Reliability. 2nd ed. UK: MacMillan Education LTD; 1986

[2] BS 7910:2019. Guide to the methods for assessing the acceptability of flaws in metallic structures. British Standards Institution. 2019

[3] ISO 55001:2014. Asset Management—Management Systems—Requirement, standard by International Standards Organization, 1st edn. 2014

[4] The Energy Institute. Guidelines for the Avoidance of Vibration Induced Fatigue Failure in Process Pipework. 2nd ed. London: The Energy Institute; 2008

[5] BS 7608:1993. Code of Practice for Fatigue Design and Assessment of Steel Structures. British Standards Institution. 1993

[6] ASME. ASME Boiler and Pressure Vessel Code Section VIII—Rules for Construction of Pressure Vessels, Division 2—Alternative Rules. ASME. 2021

[7] BS EN 13445-3:2021. Unfired Pressure Vessels—Part 3: Design. British Standards Institution. 2021

[8] Law AM, David Kelton W. Simulation Modelling and Analysis. 2nd ed. USA: McGraw-Hill; 1991

[9] Sokolov YF, Leonova OV. A Vibration Fatigue Benchmark of Fatigue Analysis Standards, ResearchGate online publication 324645552. 2018

[10] Sokolov Y. A Critical Review of ASME Weld Detail Fatigue Analysis Methods. ResearchGate online publication 258278466. 2013

[11] Bathias C, Paris PC. Gigacycle Fatigue in Mechanical Practice. USA: Marcel Dekker; 2005

[12] Bathias C. Fatigue Limit in Metals. UK: ISTE Ltd; 2014

[13] Polak TA, Pande C. Engineering Measurements Methods and Intrinsic Errors. UK: Professional Engineering Publishing Ltd; 1999

[14] Wirsching PH, Paez TL, Ortiz K. Random Vibrations Theory and Practice. USA: John Wiley & Sons Inc; 1995

[15] Somerville G, editor. The Design Life of Structures. UK: Blackie & Son Ltd; 1992

[16] Kaley LC. API RP 581 Risk-Based Inspection Methodology—Documenting and Demonstrating the Thinning Probability of Failure Calculations. Third ed. Savannah, GA: Trinity Bridge LLC; 2014

[17] DnV-RP-G101. Risk Based Inspection of Offshore Topsides Static Mechanical Equipment. Recommended Practice by Det Norske Veritas. 2010

[18] DnV-RP-G103. Non-Intrusive Inspection' Recommended Practice by Det Norske Veritas. 2011

[19] Gumbel EJ. Statistics of Extremes. USA: Echo Point Books & Media; 2013

[20] HSE RR 016. Guidelines for use of statistics for analysis of sample inspection of corrosion. TWI Limited Research Report for Health and Safety Executive UK. 2002